Recht und Verfassung in Afrika – Law and Constitution in Africa

Band/Volume 30

Oliver C. Ruppel/Holger Dix (eds.)

Roadmap for sustainable biofuels in southern Africa

Regulatory frameworks for improved development?

Konrad Adenauer Stiftung

DROP PROGRAMME

BIOFUELS research chair

WATER INSTITUTE

Nomos

The Deutsche Nationalbibliothek lists this publication in the Deutsche Nationalbibliografie; detailed bibliographic data are available on the Internet at http://dnb.d-nb.de

ISBN 978-3-8487-3306-4 (Print)
978-3-8452-7758-5 (ePDF)

British Library Cataloguing-in-Publication Data
A catalogue record for this book is available from the British Library.

ISBN 978-3-8487-3306-4 (Print)
978-3-8452-7758-5 (ePDF)

Library of Congress Cataloging-in-Publication Data
Ruppel, Oliver C. / Dix, Holger
Biofuels Regulation
Oliver C. Ruppel / Holger Dix (eds.)
268 p.
Includes bibliographic references.

ISBN 978-3-8487-3306-4 (Print)
978-3-8452-7758-5 (ePDF)

1. Edition 2017

Foreword

Prof Dr Eugene Cloete, Vice-Rector Research, Innovation and Postgraduate Studies, Stellenbosch University, South Africa

On 25 November 2015, I had the opportunity to officiate a high level interdisciplinary dialogue on sustainable biofuels in southern Africa, held at the Stellenbosch Institute for Advanced Studies (STIAS). This dialogue was unique in many ways, but especially in that it brought together experts on biofuels from various disciplines which not seldomly tend to work in isolation to each other. In this light I commend the editors of this timely publication, that they have managed to consolidate the discussions and outputs of the aforementioned dialogue into a fully-fledged book publication. This is the type of research and publication that we foster at Stellenbosch University, especially when it has potential for development and to improve the lives of the most needy in our society.

With great pleasure I see three logos of Stellenbosch University represented on the cover of this publication: The Development and Rule of Law Programme (DROP); Stellenbosch University Water Institute (SUWI) and the Stellenbosch University NRF Biofuels Research Chair. These entities at Stellenbosch University have made significant impact in terms of research and innovation, not only in South Africa, but on the continent as a whole. The fact that these three logos are accompanied by the one of Konrad-Adenauer-Stiftung or as we call in here in South Africa, the Konrad-Adenauer-Foundation underlines a long-standing and sustainable relationship between our university and this German political foundation, which has done so much good over the years.

I whole-heartedly recommend this publication as an example of high-level interdisciplinary research and am confident that it will foster the dialogue on biofuels in southern Africa and beyond. Moreover, it is my sincere hope that it will contribute to more innovative responses for development that contribute to raise the potential on sustainable energy on the continent. Regulatory frameworks are critical when it comes to finding economical responses to social challenges. In this light, policy and science are urgently called on to further their communication in order to have deeper impact in future.

Preface

Dr Holger Dix, Konrad-Adenauer-Foundation, Resident Representative, South Africa

Prof Dr Oliver C Ruppel, Konrad-Adenauer-Foundation, Director Climate Policy and Energy Security Programme for Sub-Saharan Africa; Professor of Law Development and Rule of Law Programme (DROP), Stellenbosch University, South Africa

The finite nature of fossil raw materials and the necessity of climate protection call for a worldwide reform of energy and resource policies to promote the utilisation of renewable, natural resources. The UN Paris Convention on Climate Change of December 2015 reconfirmed the associated challenges. The limitation of global warming to below 2 degrees, as contained in the convention, can only be adhered to if the emission of greenhouse gases is drastically reduced. To this end, we need not only an increase in energy efficiency and a decrease in energy and resource utilisation, but also an economy that utilises fewer fossil raw materials and that relies more on renewable, environmentally friendly and natural resources. Environmental and energy policies must now strive to promote a viable energy mix that will allow quality of life, that is guided by criteria of sustainability and that at the same time guarantees competitiveness of economies. In Germany, the realisation that energy supply and environmental protection must go hand in hand has led to energy reform that enjoys international recognition. But, also outside Germany, energy supply concepts and legal frameworks are being developed in which the importance of renewable energies is gaining significance. In this process, a bio-based economy is often seen as key for a sustainable and viable economy and as an essential prerequisite for the conservation of the environment. The particular significance of bioenergy results from its versatility and the high-energy density of biomass. Biomass can be made available as a solid, liquid or gaseous energy source; it can replace fossil fuels; and it does not depend on fluctuating sources such as wind or sun. Bioenergy can be stored and can provide sufficient power to cover the base load.

The expectations concerning energy reform, however, vary greatly depending on the respective local conditions. In the global South, climate change is still considered by some citizens and political decision makers as a luxury problem of the global North. The democratic support required for a change of policy is in many instances still insufficient. Yet it is evident that the poorer regions of the world will be particularly affected by climate change, because of their lesser ability to adapt. In countries such as South Africa that have a growing energy need and that consider their ample coal reserves as a competitive advantage, the possible economic effects of energy reform form the subject of controversial discussions. In regions facing problems of food supply, it is rightfully pointed out that the increased utilisation of bioenergy may not be in competition with food security, since the area of land available for the cultivation of food is limited. The debate around sustainable utilisation and protection of resources always includes the necessary consideration of policy direction towards medium- or long-term goals and the consideration of possible conflicts of interest. In the case of bioenergy, the possible conflicts of interest could also include the use of the scarce resource of water in the cultivation of renewable energy sources or in the conservation of biodiversity.

The Development and Rule of Law Programme (DROP) of the University of Stellenbosch and the Konrad-Adenauer-Foundation addressed these issues in a workshop in November 2015. The objective of the workshop was to allow for an exchange of information and experiences in the field of international policies and regulations concerning bioenergy; to analyse socioeconomic factors concerning the utilisation of bioenergy; to conduct a comparison of the legal foundations of bioenergy in selected countries; and to present examples of best practice. The excellent contributions by workshop participants prompted the organisers to launch this publication. It is our hope that it will draw wide attention and lead to numerous ideas for the further utilisation of bioenergy in southern Africa.

January 2017 *The Editors*

Table of contents

PART I:
BIOFUELS – THE SCIENCE BASE

The science base for bioenergy

Prof Dr Willem H van Zyl, NRF SARChI Research Chair: Biofuels, Stellenbosch University, South Africa

Prof Dr Johann F Görgens, Department of Process Engineering, Stellenbosch University, South Africa

Prof Dr Alan C Brent, Associate Director of the Centre for Renewable and Sustainable Energy Studies (CRSES), Stellenbosch University, South Africa

A. Introduction

I. Current role of bioenergy in Africa

The size of Africa is often underestimated – the entirety of the USA, all of India and China, and almost all of Europe can fit into Africa's landmass.[1] Of the 3,022 million hectares of land in Africa, an estimated 840 million hectares are arable land, compared to 890 million hectares in the Latin Americas. However, only about 27% of the arable land in Africa is utilised, leaving about 600 million hectares unused.[2] At the same time, this huge continent has been poverty-stricken, with almost 50% of the population living on less than $1.25 per day.[3] More than 620 million people do not have access to electricity, and nearly 730 million people are still dependent on firewood for daily household energy requirements. When excluding South Africa, 90% of the 280 million tonnes of oil equivalent solid biomass that is used in sub-Saharan Africa is for household cooking and heating. Although the roll-out of electricity in Africa is recommended more than 500 million people will remain without it as a result of rapid

1 Krause (2010).
2 Ejigu (2008).
3 UN (2012).

population growth,[4] and specifically rapid urban growth accompanied by massive urban poverty and many other social problems.[5]

Against the backdrop of poverty and lack of electricity in sub-Saharan Africa, about a third of the region is covered by forest, and the total forest biomass was estimated to be a staggering 130 billion tonnes in 2010. However, less than 1% should be used annually so as to not promote deforestation.[6] It is not surprising that about 48% of Africa's primary energy supply is from biomass and the bulk of the rest from crude oil (22%), coal (14%, primarily in South Africa), and natural gas (14%), while other renewables and hydropower make up the final portion (2%). The bulk of biomass is used in the form of firewood, which is not only hazardous, but also rather inefficient.[7] In urban areas, charcoal is a popular fuel option, because of its higher energy density and the ease with which it is transported, stored, and distributed. However, the bulk is still home-produced in traditional earth-mound kilns that have a conversion efficiency of 8 to 12%, while industrial kilns are more than twice as efficient.[8] However, although the average household income may rise in some countries in sub-Saharan Africa, changes in the use of household fuel will be slow, particularly in rural areas.

The preferred biomass sources, and the associated inefficiencies, result in significant deforestation,[9] which is a major environmental concern.[10] At the same time, significant amounts of biomass residues and invasive plants responsible for bush encroachment remain underutilised in terms of energy production.[11] The use of modern bioenergy – which constitutes a switch away from deforestation towards sustainable biomass sources – will not only address these inefficiencies and changes of fuel mixes, but can play a significant role in addressing socioeconomic challenges in sub-Saharan Africa.

4 IEA (2014).
5 UN-Habitat (2014).
6 IEA (2014).
7 IRENA (2015).
8 IEA (2014).
9 Allen & Barnes (1985).
10 Chidumayo & Gumbo (2013); Subedi et al. (2014).
11 Smith et al. (2015); Mugido et al. (2014).

II. Role for modern bioenergy in Africa

Looking towards the future, modern bioenergy holds the promise not only to use biomass energy more efficiently, but also to provide economic, social and environmental benefits. Within the African context, modern bioenergy could include both traditional biofuels, such as biogas, bioethanol, biodiesel and even more efficient charcoal production, and the roll-out of advanced lignocellulosic conversion technologies to cellulosic ethanol, high density pyrolysis oils, char and gases, to syngas for the production of synthetic fuels (Figure 1). The particular value of advanced lignocellulosic conversion technologies is that they provide options for combining the production of traditional and advanced biofuels, and optimise biomass utilisation. These technologies also facilitate the integration of bioenergy production with existing bio-based industries, such as paper and pulp, and sugar industries, waste streams valorisation, and electricity production for local decentralised grids. The implementation of modern bioenergy also requires a simultaneous switch to sustainable biomass sources, without negatively affecting key social services, such as food production or the preservation of the natural environment.

III. Potential for modern bioenergy production in Africa

Projections for bioethanol production potential in sub-Saharan Africa countries vary, from a modest 2.4 billion litres from existing capacity and feedstocks available,[12] to a potential 170 billion litres projected for 2030.[13] For the larger part of sub-Saharan Africa there is a significant gap between the current and the potential production of liquid biofuels. Bioethanol produced from sugarcane presents an interesting case. During the 1980s several successful biofuels plants, using primarily sugarcane molasses as feedstock, were commissioned in Zimbabwe, Malawi and Kenya. The Zimbabwean Triangle Ethanol plant produced about 40 million litres annually, the Dwangu Estate Plant in Malawi 15 to 20 million litres annually, and the Muhoroni Plant in Kenya about 15 million litres,[14] allowing the blending of 10 to 15% into petrol in these countries. Unfortunately, be-

12 Amigun et al. (2011); see also http://www.epasa.org.za/about-us.html, accessed 24 July 2016.
13 IRENA (2015).
14 Amigun et al. (2011); Deenanathi et al. (2012).

cause of droughts, financial crises, and the lack of governmental support, production declined and some of these plants have even closed down. The ethanol production in Africa was 637 million litres in 2007, of which 388 million litres came from South Africa, with the bulk of synthetic ethanol produced by Sasol.[15]

The total sugar production in sub-Saharan Africa was 10.68 million tonnes in 2008, versus 4.78 million tonnes in southern Africa, and 2.80 million tonnes in South Africa alone.[16] If considering a sugar production of about 10 million tonnes in sub-Saharan Africa, and applying the Brazilian model of approximately 50% sugar to 50% ethanol production, sub-Saharan Africa can produce 3 billion litres of ethanol, which is tenfold more than the current bioethanol production from primarily sugarcane molasses. As part of a larger COMPETE programme, the University of KwaZulu-Natal has shown that an estimated 6 million hectares of land, suitable for sugarcane production, are available in Angola, Malawi, Mozambique, Tanzania, Zambia and Zimbabwe, above that already providing primarily for food production and conservation.[17] Considering the conservative yield of 65 tonnes per hectare of sugarcane produced under relatively dry conditions in South Africa, the 6 million hectares could yield roughly 400 million tonnes of cane per annum, twentyfold more than that produced by the South African sugarcane industry. If first and second generation technologies are used, this could potentially yield 35 to 49 GL of biofuels, depending on the amount of sugarcane used for biofuels production, which represents about 20% of Africa's current total petroleum consumption.[18] Africa thus has the potential of becoming another Brazil in terms of biofuels production, and South Africa can play a key role of realising this potential as a major technology provider.

IV. Socioeconomic benefits of modern bioenergy

Modern bioenergy promises to provide economic, social and environmental benefits to particularly sub-Saharan Africa at large. Because biomass production is inevitably linked to the farming community, bioenergy production can improve the social well-being of rural communities. Reducing

---

15 Munyinda et al. (2011); Ambali et al. (2011); see also http://www.epasa.org.za/about-us.html, accessed 24 July 2016.

16 Hassan (2008).

17 Watson (2011).

18 Van Zyl et al. (2011b).

the time-consuming collection of firewood, by particularly women, could mean better caring for children and their education. At the same time it could reduce health hazards associated with in-house smoke inhalation.[19]

Modern bioenergy could also provide additional sources of income and decentralised energy (heat, electricity and biofuels for transport) to local communities, particularly those in land-locked countries. This could overcome the 'double penalty' of the higher cost of importing fertilisers and then transporting goods to seaports. Food losses because of a lack of refrigeration and storage energy could also be reduced, which would help to overcome food insecurity and malnutrition. Furthermore, creating other markets for biomass parallel to food markets, and thus making food production less vulnerable to the vicissitudes of international markets, would also aid food security.[20]

Modern bioenergy would be more efficient than the use of traditional firewood and charcoal produced in traditional earth-mound kilns, because less carbon is released, leading to climate change mitigation.[21] The production of bioenergy does not preclude the production of other renewable energies, but can help with job creation in rural communities. It is important that biomass for bioenergy does not fall into the same trap as large-scale raw material exports, as in the case of crude oil and precious minerals. Bioenergy production should first focus on empowering local communities, by directing attention on smallholder production and processing; thus empowering farmers by generating new income streams that can help with rural economic transformation. Domestic markets should preferably be developed, rather than international markets. Except for advanced lignocellulosic and similar technologies, more traditional modern bioenergy technologies (for example biogas and biodiesel for households and farming equipment) are simple and easily transferable, and can be operated at village level.[22]

V. Focusing on South Africa

South Africa's size is approximately 1.2 million square kilometres, with a large portion being considered semi-desert. Owing to the aridity of the

19 Ejigu (2008).
20 Lynd et al. (2015).
21 Ejigu (2008).
22 (Ibid.).

land, only 12% can be used for crop production, and only 22% of this is considered high-potential land. Agricultural activities range from intensive crop production and mixed farming in winter rainfall and high summer rainfall areas, to cattle ranching in the bushveld and sheep farming in the arid regions. Maize is most widely grown, followed by wheat, sugarcane, and sunflowers.[23] About 1% of the total land area is devoted to forestry in the form of pine (52%) and Eucalyptus (39%) plantations in the provinces of Mpumalanga (40%) and KwaZulu-Natal (39%).[24]

VI. An appropriate framework for implementation in Africa

The demonstrated potential of modern bioenergy to contribute to energy supply in the African context warrants the consideration of an appropriate framework for implementation on the continent, in particular owing to potential synergistic benefits between energy supply, food production, and poverty alleviation.[25] Whereas developed countries pursue bioenergy options primarily because of potential environmental benefits, the drivers in the African context are different, taking a more holistic view to desired socioeconomic development and sustainability. Particular aspects of sustainability that carry higher priority in the African context would be socio-economic development, overcoming energy poverty, and sustainable use of bio-based natural resources.[26] Energy poverty in Africa is not only defined by lack of energy access, but more importantly also by the unaffordability of energy services, such as electricity and imported fossil fuels.

It is important to note that the sustainability and desirability of bioenergy cannot be determined holistically, since many approaches exist for its production. A multiplicity of potential biomass feedstock options, available conversion technologies, and associated end-products lead to a large number of potential value chains for bioenergy production (see Section C). A framework is required to evaluate each of these value chains in terms of economic, environmental and social indicators of sustainability, and thereby to distinguish those that are preferred from those that are not.[27] Several approaches to bioenergy can achieve the desired benefits in

23 GCIS (2015).
24 DAFF (2011).
25 Mirzabaev et al. (2014).
26 Ahlborg & Hammar (2014); Lucas et al. (2015).
27 El Fadel et al. (2013); Mafakheri & Nasiri (2014); Schut et al. (2014); Sarrafa et al. (2013); Schut & Florin (2015).

all aspects of sustainability, while at the same time it is always possible to define less desirable approaches to bioenergy production that will do more harm than good.

The remainder of the present chapter will review the diversity of biomass feedstocks and conversion technologies available for implementation in the African context, sketching the diversity in potential value chains for bioenergy production. The assessment of particular value chains against the three key pillars of sustainability, namely economic, environmental and social impacts, is proposed as a means of identifying preferred options, as demonstrated with a case study from South Africa.

B. Overview of different types of biomass resources for bioenergy production in Africa

Biomass for bioenergy production can be classified according to its method of production as well as its chemical composition. Firstly, bioenergy can be produced from biomass residues/wastes from existing activities, or from dedicated/new bioenergy crops, grown specifically for energy production. Examples of residues/wastes include: agricultural residues, such as straw, sugarcane leaves, forestry residues, and corn stover; residues from biomass processing, such as fruit or milling residues from agroprocessing of food, sugarcane bagasse, and fibrous residues from sawmills or biomass pulping; municipal and domestic wastes, such as municipal solid waste, and sewage sludge; and plant materials obtained from clearing activities to protect indigenous plants, biodiversity and agricultural lands from unwanted plant growth, such as invasive alien plants and encroaching bush. Several of these residues have existing (low-value) applications, while others are simply discarded by the simplest means, sometimes resulting in further land degradation through landfilling.[28]

Alternatively, bioenergy crops can be cultivated through appropriate agricultural methods, to produce crops like sugarcane; sweet sorghum; sugarbeet; grasses such as, switchgrass, Napier grass, giant bamboo and Miscanthus; starch-rich grains; cassava; and oil-bearing seeds/plants. However, it is important that these crops are not considered invasive species.[29] Starch-rich grains can either be existing food crops, such as barley, wheat or corn, or non-food crops, such as sorghum grain or triticale. Oil-

28 Ho et al. (2014); Batidzirai et al. (2016); Pan et al. (2015).
29 See the chapter by Blanchard et al. in this publication.

bearing seeds/plants include: tobacco, canola, rapeseed, groundnuts, sunflower and soybeans. The production of algae as a feedstock for bioenergy production is typically motivated by the potential to utilise CO_2 produced from a range of existing industrial processes as carbon source, as well as the potential for integration of algae production with waste water treatments.

The biomass examples listed above can be further classified according to chemical composition, which determines what kinds of bioenergy processing, or end-products, the various biomass types can be utilised for. Energy crops are typically cultivated for a particular highly accessible biomaterial for energy production.[30] Sugarcane, sweet sorghum and sugarbeet, for example, contain significant amounts of water-soluble sugars, which can be readily extracted and converted into alcohols, such as ethanol, as a biofuels. Starch-rich crops, like grains and cassava, are typically utilised for the production of fermentable sugars, by selective hydrolysis of starches through applications of heat and enzymes. Feedstocks such as oilseeds from canola, rapeseed or soybeans, are processed through pressing and solvent extraction, to obtain vegetable oils suitable for the production of transport fuels, such as biodiesel and jetfuel. Algal biomass is typically selected for oils/lipids or starch, although some species (such as macroalgae) do not contain significant quantities of these. The particular chemical composition of these examples of readily accessible forms of plant biomass makes them suitable for fairly uncomplicated means of conversion to products like biofuels.

The majority of wastes/residues available for bioenergy production come in the form of fibrous plant biomass, also known as lignocellulose.[31] Examples of lignocellulose-containing materials include: straw, bagasse, forestry residues, invasive plants, encroaching bush, grasses, firewood, as well as the residues of energy crops after sugar/starch/oil extraction. Lignocellulose comprises cellulose, hemicellulose, lignin, ash, and other minor components. Lignocellulose is by far the most abundant form of biomass in nature, while also being the most generic of biomass sources for bioenergy production, since it is suitable for conversion to all types of bioenergy, including: thermal energy (heat), electricity, and solid/liquid biofuels for further applications.[32]

---

30 Ho et al. (2014); IEA-RETD (2015).

31 (Ibid.).

32 Görgens et al. (2014); Ho et al. (2014); Batidzirai et al. (2016); IEA-RETD (2015).

The availability of the various types of biomass for energy production depends on a multiplicity of factors, if both existing and potential future land uses are considered. In the South African context, the so-called Bio-Energy Atlas was developed by the Department of Science and Technology, giving indications of the geographic distribution of existing and potential future sources of biomass for bioenergy production.[33] Similar efforts have been made to develop biomass availability models for Mozambique and other southern African countries.[34]

Table 1: Lignocellulosic biomass availability in South Africa (in units of Mt/a)[35]

1. Agricultural residues	
Maize stover	31.1 Mt/a
Wheat straw	2.4 Mt/a
Grain sorghum	0.2 Mt/a
Rest of agricultural residues	2.5 Mt/a
Agricultural subtotal	**36.2 Mt/a**
2. Sugarcane residues	
Field residues	5.1 Mt/a
Bagasse	5.3 Mt/a
Sugarcane subtotal	**10.4 Mt/a**
3. Forest industry residues	
Plantation residues	6.7 Mt/a
Paper and pulp mill residues	0.7 Mt/a
Black liquor	1.5 Mt/a
Sawmill residues	3.1 Mt/a
Forest industry subtotal	**12.0 Mt/a**
Invasive plant species	11.3 Mt/a
Fuelwood	14.0 Mt/a
Total, annual basis	**83.9 Mt/a**

33 Hugo (2015).
34 Batidzirai et al. (2016); Schut et al. (2014); Schut and Florin (2015).
35 Hugo (2015).

The lignocellulosic biomass available in South Africa is estimated at about 84 million tonnes per annum (see Table 1). This correlates well with the estimations.[36] When advanced second-generation biofuel technologies come to fruition, and 50% of the residual lignocellulosic biomass (almost 30 million tonnes on an annual basis) is used, biofuels could play a significant role in South Africa's transport fuel future.[37] If integrated biomass-to-liquid (BtL) fuels, from both biochemical and thermochemical processes, is used, the contribution from biofuels could represent 45% from agricultural residues, 15% from forestry residues, and 10% from sugarcane residues – amounting to 73% of the total petrol and diesel use in 2014.[38] If 30% is added from invasive plants and firewood (limited quantities depending on seed banks for invasive species), biofuels could provide the equivalent of South Africa's petrol and diesel usage.

C. Available conversion technologies for bioenergy production, leading to multiple alternative value chains

Depending on the intended end-application and demand, a wide range of conversion technologies can be applied to obtain valuable energy products from primary biomass resources. These conversion methods can be defined as chemical, biochemical or thermochemical methods, and can be applied to produce thermal energy and/or biofuels (liquid, solid or gaseous) as valuable energy end-products, including: electricity, biofuels for transportation, cooking fuels, and process heat. The various alternatives in terms of biomass feedstock, and associated options for conversion to energy products, result in a multiplicity of potential bioenergy value chains, with each value chain defined by its unique combination of a particular biomass feedstock with an appropriate conversion technology, leading to a particular bioenergy end-product. Examples of the resulting 'superstructures' are presented in Figures 1 and 2, indicating the multiple possible routes for bioenergy production.

36 Lynd et al. (2003).

37 (Ibid.).

38 SAPIA (2014).

Figure 1: Biological and thermochemical routes for complete conversion of different agricultural and forestry feedstocks to bioenergy

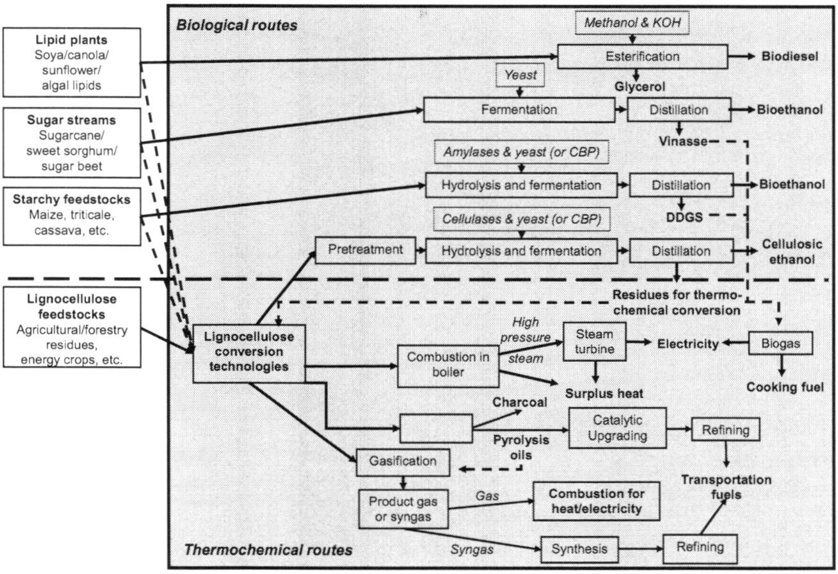

Electricity is produced by the combustion of either primary biomass feedstock, or the liquid/solid/gaseous biofuels produced from these. Combustion typically occurs in a steam boiler, gas turbine or gas engine, resulting in mechanical energy that drives the generator responsible for electricity production. Solid biofuels are produced by mechanical processing, torrefaction or the pyrolysis of biomass materials, and are useful for the production of both electricity and thermal energy (heat in various forms) as final energy services. Gaseous biofuels are produced primarily by gasification or anaerobic digestion, and can be applied for electricity or thermal energy production, while some are also utilised directly in transportation, such as biogas-driven cars and buses. A wide range of technologies is available for the production of liquid biofuels, which can be applied as feedstock for electricity, heat production, or as transportation fuels. Approaches to liquid biofuels production include: fermentation to ethanol/butanol, pyrolysis to liquids followed by the chemical upgrading of these, and gasification to syngas followed by synthesis.[39] Biodiesel production is typically considered as a chemical

39 Anex et al. (2010).

conversion technology, where the chemical components of lipid-feedstock are converted through esterification to produce biodiesel.

Figure 2: Many multiple bioenergy value chains exist, depending on feedstock, choice of conversion technology and expected products. This is demonstrated through the potential thermochemical technologies for conversion of lignocellulosic plant biomass[40]

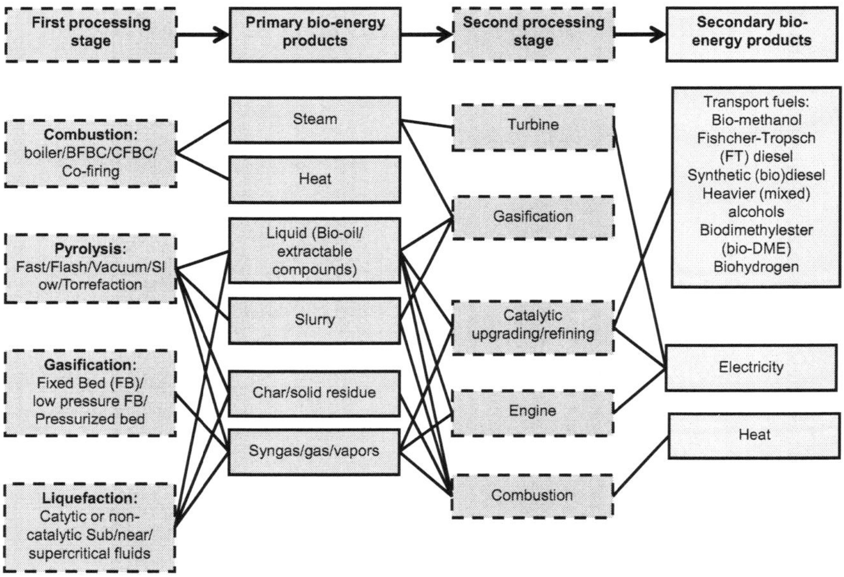

Biological conversion technologies include: anaerobic digestion, hydrolysis-fermentation, and gasification followed by fermentation – each with unique feedstock requirements and potentials.[41] Anaerobic digestion (AD) for biogas production is applied to feedstock mixtures with appropriate carbon/nitrogen ratios of chemical composition, and being sufficiently digestible to allow the process to work efficiently. Hydrolysis–fermentation is applied to both starches and lignocellulose for the production of alcohols; the process can be done without hydrolysis in the case of feedstock with high quantities of water-soluble sugars. Gasification-fermentation has been proposed as an alternative means of

40 Source: Görgens et al. (2014).
41 Görgens et al. (2014).

converting lignocellulose and municipal solid wastes to ethanol.[42] Thermochemical technologies include: combustion, gasification, pyrolysis, and direct liquefaction, often combined with subsequent processing to achieve the desired final energy products, as per Figure 2.[43]

The properties of a particular biomass feedstock have an impact on the type of processing that would be suitable for the conversion to bioenergy products, or, conversely, different bioenergy technologies have different feedstock requirements, to make them suitable for a particular conversion process.[44] Feedstock materials that contain significant quantities of water-soluble sugars or starches are typically processed via fermentation, because this type of feedstock is highly suitable for this process. Food wastes, animal manure, waste waters, and sewage sludges are typical feedstock for anaerobic digestion,[45] since these are more amenable to microbial degradation (also referred to as digestibility), while lignocellulose can be utilised in limited quantities in mixtures with other feedstock, owing to its limited digestibility.[46]

With lignocellulose being the most generic of the different bioenergy feedstock materials, it can be applied in virtually all the biological or thermochemical conversion processes, resulting in multiple processes competing for access to the same feedstock.[47] Firstly, lignocelluloses are generally suitable and preferred for application in electricity or thermal energy production, although intermediate upgrading through thermochemical or mechanical processing can be applied to improve its fuel quality for these applications.[48] An alternative application is the biological conversion of lignocellulose to ethanol via pretreatment–hydrolysis–fermentation, which has been compared to alternative thermochemical conversion routes, such as pyrolysis-upgrading or gasification-synthesis,[49] all of which result in liquid biofuels for transportation. Whereas the biological conversion routes would prefer lignocelluloses with higher contents of celluloses and hemicelluloses, preferably with increased digesti-

42 Dwivedi et al. (2009).
43 Görgens et al. (2014); Panwar et al. (2012).
44 Meincken & Tyhoda (2014).
45 Zhang et al. (2013).
46 Sawatdeenarunat et al. (2015).
47 Görgens et al. (2014).
48 Meincken and Tyhoda (2014).
49 Anex et al. (2010); Leibbrandt et al. (2011).

bility, the thermochemical routes would prefer lignocelluloses with higher lignin and energy contents.[50]

The number of possible value chains is further expanded when considering location-specific aspects of bioenergy implementation. For example, integrating bioenergy production with, or annexing it to, existing industrial facilities is widely proposed to be a more economically viable means of its production.[51] Annexing bioenergy production to existing sugar mills, grain mills, pulp mills, or electricity plants is likely to provide cost and efficiency benefits that will reduce the production cost of bioenergy. One example which is particularly relevant to the African context is the potential to integrate modern bioenergy into postharvest processing of food crops in rural areas. The resulting improvements in energy availability enable mechanisation of these processes, with demonstrated benefits in terms of reduced food losses, and economic viability of food processing.[52]

The multiplicity of options for bioenergy production requires the definition of a common framework for the evaluation or assessment of individual options. The aim of such a framework is to define a substantial set of criteria for the assessment of individual bioenergy value chains, and through that to enable the identification of preferred or desired options, with the elimination of undesired options.

D. *Sustainability indicators and comprehensive quantification of bioenergy options, as a means to identify preferred approaches*

Various international efforts have aimed to identify preferred bioenergy production options on the basis of the assessment of individual options against the three pillars of sustainability, namely: environmental, social, and economic impacts. Organisations, such as the World Wide Fund for Nature (WWF) and the Food and Agriculture Organization of the United Nations (FAO), have defined sustainability standards for bioenergy, which typically would combine a set of minimum criteria to be met in each of the three pillars of sustainability.[53] These sustainability standards allow for the (quantitative) assessment of individual bioenergy production options, as per the multiplicity of alternatives created by localised biomass availabil-

50 Meincken & Tyhoda (2014).
51 van Zyl et al. (2011a); Chimphango & Görgens (2015); IEA-RETD (2015).
52 Chimphango & Görgens (2015); Padi (2016); Lynd & Woods (2011).
53 GBEP (2011); Fritsche et al. (2006).

ity, conversion options, and energy end-user needs, to identify preferred and non-preferred options. The impacts of bioenergy production are determined with a life-cycle and systems approach, thus taking the full extent of impacts into consideration.[54] For example, an environmental impact assessment would include an evaluation of land-use criteria and availability, loss of biodiversity and deforestation, the effect of emissions on the environment, soil erosion and degradation, and water-use and -contamination. Social impacts would include aspects such as: education, job creation, livelihoods, worker conditions, and competition with productive capacity for food and other societal needs from available biomass. Economic impacts would consider not only the viability of bioenergy as an investment, but also the wider economic benefits, for example the balance of payments versus crude oil imports, energy security, and macroeconomic development. As per the example of greenhouse gas (GHG) emissions, it is important that these assessments be done in a quantitative rather than qualitative manner. For example, it has been shown that the corn-ethanol industry in the United States of America, for most of its existence, has not contributed significantly to GHG reductions, although improved monitoring and reporting have in recent years resulted in a measurable GHG emissions benefit.[55]

Implied in the social impact assessment of bioenergy production is the recognition of societal benefits and priorities for the utilisation of available plant biomass in a particular region. For example, Stafford and Brent[56] evaluated the different bioenergy options to utilise invasive species on the Agulhas Plains of the Western Cape Province of South Africa, with the aim to maximise the socioeconomic benefits for the local communities. Various hierarchies of priorities for biomass applications, according to the priorities of societal needs, have been defined (see Figure 3), with the intention that higher societal priorities must not be impacted negatively by the production of bioenergy.[57] Higher priority applications typically carry greater economic value per unit of carbon utilised from the biomass feedstock, but this is not necessarily the case, and is also not a fail-safe to ensure that quantitative sustainability assessments will prioritise these. Furthermore, increased demand for feedstock for lower-priority applications, such as bioenergy, may also increase the feedstock cost and end-product

54 (Ibid.).
55 Gallagher et al. (2016).
56 Stafford & Brent (2011).
57 (Ibid.).

cost for higher priority applications, such as food or animal feed – which is an undesirable outcome.

Figure 3: The biomass value addition pyramid The added value is the highest at the top of the pyramid, whereas the volume of biomass needed is the highest at the bottom of the pyramid.[58]

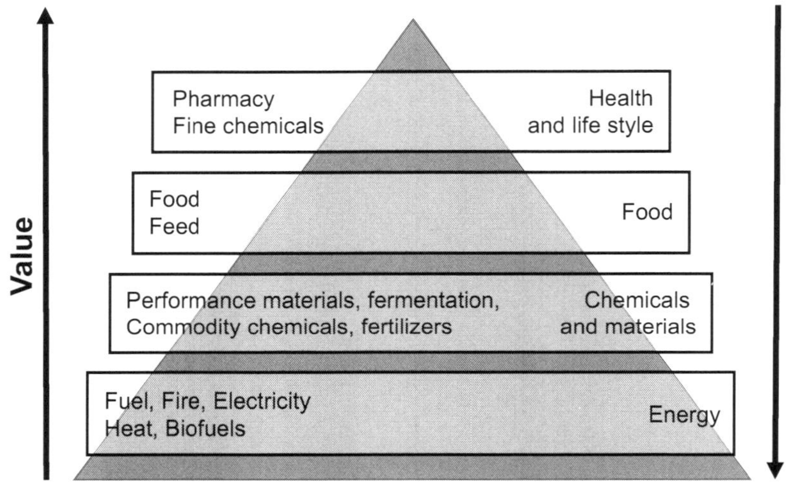

Given the multiplicity of competing applications for a limited amount of biomass, not only for energy production, but also for higher value applications such as health (pharmaceuticals), food, animal feed, and the manufacture of chemicals and materials, it is clear that substantial competition exists for biomass feedstock as part of the natural resources of a particular region.[59] The resulting conflicts are best addressed through: (i) an assessment of alternative applications against sustainability criteria and standards; and (ii) the recognition of the increased societal priority for particular applications which should be given preference over bioenergy applications. Stafford and Brent[60] provide an analytical framework which shows the cycle of active learning and research for the assessment, management and monitoring of bioenergy interventions (see Figure 4). For such a framework to be effective, different decision support systems and tools are used at each stage to aid in the comprehensive assessment,

58 Source: Eickhout (2012).
59 von Maltitz et al. (2009).
60 Stafford & Brent (2011).

management and monitoring of bioenergy interventions. Given the limited amounts of land and other natural resources available, it is essential that there is policy coherence to guide developments in bioenergy production and biomass application, to ensure that these sectors are developed in a sustainable manner by using such a framework.

Figure 4: An analytical framework for the comprehensive sustainability assessment, management and monitoring of bioenergy interventions[61]

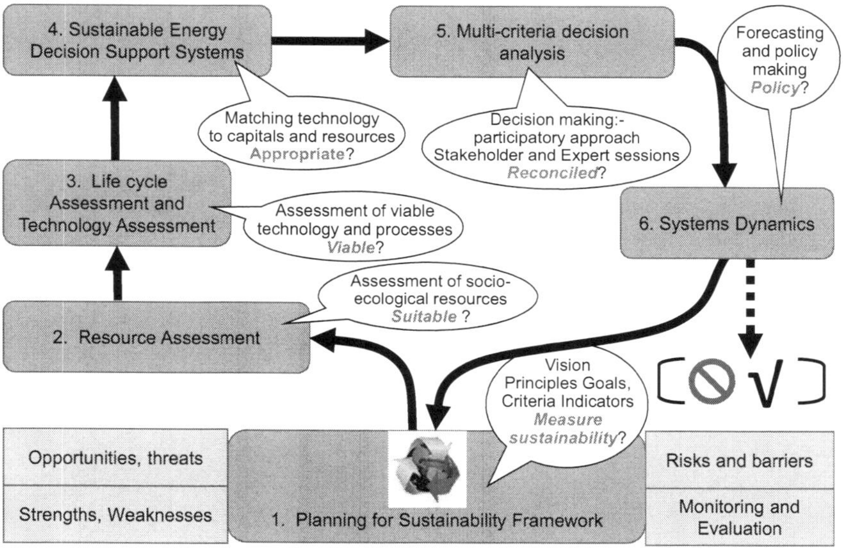

E. *Economic viability, in particular affordability, for end-users in the African context*

Unaffordability is a key constraint to energy access, which is also carried over to bioenergy, in the developing world.[62] For example, electricity access for the population of Zambia is determined by affordability, rather than availability.[63] Biomass as feedstock for bioenergy production is typically more expensive than fossil fuel-based equivalents, owing to a lower energy density and increases in transport costs.

---

61 Source: Stafford & Brent (2011).
62 Winkler et al. (2011).
63 Haanyika (2008).

Of the bioenergy options, crop-based bioenergy not only carries concerns about impacts on food production (see section D), but also presents more expensive bio-based feedstock. Wastes and residues from higher value applications in the hierarchy are thus generally preferred, owing to limited potential for impacts on higher value applications, and lower costs per unit.[64] Waste-based feedstock also carry other potential benefits, such as avoided disposal costs, which result in a negative feedstock cost, or revenue, for the bioenergy production facility, and avoided environmental impacts of the disposal of these wastes.

In some land-locked African countries the issues of energy-poverty are exacerbated by the importation of fossil-based electricity and fuels.[65] The local production of bioenergy thus has an increased number of opportunities to be economically competitive with fossil fuels. In some cases bioenergy is cheaper than the imported fossil-based energy sources, thereby addressing energy-poverty.[66]

F. Case study from the Western Cape Province of South Africa: Triticale-ethanol product

The production of ethanol from triticale as an energy crop in the Western Cape Province of South Africa is under consideration owing to various potential social and environmental benefits. Firstly, the agricultural potential of the dryland areas of the Western Cape is limited as to the types of feedstock that can be grown under rain-fed conditions, if the aim is to avoid the use of irrigated water for energy crop production. Small grains, such as wheat, barley and triticale, are well suited to these conditions. Furthermore, the land under cultivation for wheat and barley production in the Western Cape has been shrinking continuously over the past 20 years, owing to globalisation and changing market conditions. Only the highest-yielding lands remain under wheat production, thus making marginal lands – which are no longer economically suitable for wheat production – available for the cultivation of triticale as a non-food feedstock for ethanol. Triticale requires lower agricultural inputs than wheat, while also delivering higher yields on these marginal lands. This gives triticale a competitive advantage over wheat in agriculture. The conversion of triticale to

64 Ho et al. (2014); Sarkar et al. (2012); Batidzirai et al. (2016).

65 Arvis et al. (2007).

66 Amigun et al. (2008).

ethanol follows a very similar process to the well-established industrial processes for wheat and corn, while the increases in small-grains agriculture through triticale production will positively impact on job creation.

Two aspects of the manner in which triticale-ethanol production is implemented require careful consideration to ensure sustainability. The first of these is environmental impacts, in particular the GHG emissions associated with the life cycle of this bioenergy value chain. Substantial fossil-based GHG emissions will be incurred in production, harvesting and transport of triticale, through the use of fertilisers and diesel vehicles. Furthermore, the production of ethanol from triticale requires significant external energy inputs in the form of steam and electricity, known collectively as process energy. If process energy is obtained from fossil sources, for example for coal-based steam and electricity production, then the overall life cycle GHG emissions would be larger than the GHG benefits obtained by the use of the final ethanol product to replace fossil-based gasoline. One means to address this is to utilise residual biomass, such as invasive plants from nearby clearing activities in the Western Cape Province, to provide the process energy requirements through onsite steam and electricity production. This approach will decrease the GHG emissions of the triticale-ethanol life cycle to the point where a net GHG reduction benefit, compared to the fossil-based gasoline replaced by ethanol, can be achieved.[67]

The second key consideration for sustainability is the viability of triticale as an investment, i.e. whether the economics of ethanol production from triticale will provide an attractive opportunity for investors. Various recent economic assessments have shown that the combination of proposed subsidy schemes for bioethanol in South Africa and the low crude oil and gasoline prices, along with the anticipated high cost of triticale feedstock, will result in an unattractive investment. The production cost of triticale-ethanol will be high, primarily owing to feedstock costs driven up by the present drought, while the ethanol selling price will be low, owing to low gasoline prices linked to crude oil – a situation that even the proposed subsidy schemes cannot rectify. Although the proposed triticale-ethanol project may pass various environmental and social impact assessments for sustainability, it remains economically unviable under the current conditions of drought and low crude oil prices.[68]

67 Gallagher et al. (2016).

68 du Preez (2016); GreenCape (2013).

To investigate these sustainability issues further, a system dynamics model was developed,[69] following on other studies (see step 6 in the analytical framework in Figure 4). The main drivers and areas of concern affecting the biofuels industry were identified through surveying the latest literature specific to the Western Cape Province and consulting with industry experts.[70] The following key aspects were consistently mentioned and directly influence the feasibility of biofuel production and the required compliance with the mandatory blending policy:

- *Biofuel production plant* – The type and size of plant influences building cost, resource use and production efficiency, which directly drives the capital expenditure (CAPEX) and operational expenditure (OPEX).
- *Agriculture* – The type of agricultural crop used has an impact on production as one has to adjudicate based on feedstock cost, availability and biofuel yield of the specific crop.
- *Food security* – The country's food security is dependent on the local agriculture sector, and the allocation and selection of crops to produce biofuel has to be done in a manner that will not negatively impact on food security.
- *Land resources* – Limited land area in the Western Cape Province (19% of the province's total land) is suitable for crop growth and it is important that crops grown for biofuel production do not encroach on land allocated for food crop production. The size of the planting area and the type of crop used for production directly influences the land requirements.
- *Water resources* – Producing biofuel places additional stress on water resources through increased agricultural activities, process water, and make-up steam and cooling water.
- *Energy requirements* – Biofuel (especially bioethanol) production is energy intensive and, in the midst of South Africa's energy crisis, it is necessary to consider alternative energy sources and production strategies.
- *By-products* – Depending on the production process used, the by-product of biofuel production could strengthen the business case. Dried distillers grains with solubles (DDGS) is a by-product of both bioethanol and biodiesel production processes and is commonly used

69 Jonker et al. (2015).
70 These included: Amigun et al. (2012), Nolte (2007), GreenCape (2015), and Fore et al. (2011).

as animal feed. Glycerine (used in the cosmetic and pharmaceutical industry) is also produced as a by-product in biodiesel production. The local market for glycerine is, however, saturated owing to a large global over supply, and glycerine thus constitutes a waste management challenge.

- *Biofuel sales price* – The financial feasibility of biofuel production ultimately depends on income generated from sales. The sales price is normally defined as a percentage (based on the ratio of biofuel to fossil fuel energy density) of the basic fuel price (BFP), which is determined by the crude oil price and rand-dollar exchange rate.

In addition to the influencing factors in the biofuel sector, indicators were used to assess the impact and feasibility of different biofuel production scenarios;[71] these indicators are identified and described in Table 2.

Once the drivers and indicators were identified, they could be schematically represented in order to find the points of relation with interconnecting sectors, and to serve as a visual aid to detect and expand on possible exogenous drivers and endogenous effects.

I. Causal loop modelling

In system dynamics, the visual representation of drivers and indicators is done in the form of a causal loop diagram (CLD). CLDs consist of balancing (B) and reinforcing (R) loops, as indicated in Figure 5, where a balancing loop shows two entities oppositely affecting one another, while variables in a reinforcing loop influence each other to the same extent. This is best described by means of an example. In R1 an increase in population will lead to an increase in births, which again increases the population (thus reinforcing). However, in B1 an increase in population will lead to an increase in deaths, in turn decreasing the population (thus balancing).

71 Jonker (2015).

Table 2: Modelling indicators[72]

	Indicator	Description	Units
Environmental	Emissions	Change in air emissions due to biofuel sector	kg CO_2 / year
	Land use	Agricultural land required for biofuel feedstock cultivation	Ha
	Water use	Water resources required for bio-fuel production	Litre / year
	Electricity use	Energy used to produce biofuel	kW
	Biofuel by-product	Amount of DDGS produced per year	Litre / year
Social	Employment	Additional employment creation due to biofuel production	Person
Economic	Investment into biofuel	Capital investment needed to produce biofuel	Rand
	Operational cost	The running cost of a production facility to produce biofuel	Rand / year
	Subsidy re-quired	The subsidy required per litre to attain a 15% return on assets (ROA)	Rand / litre
	Internal rate of return (IRR)	Profitability of investing into biofuel production	%
	Bottom line	The overall bottom-line cost im-plications of providing biofuel to Western Cape consumers through local production or inter-provincial imports	Rand

72 Source: Jonker (2015).

Figure 5: A causal loop diagram for the biofuel sector in South Africa[73]

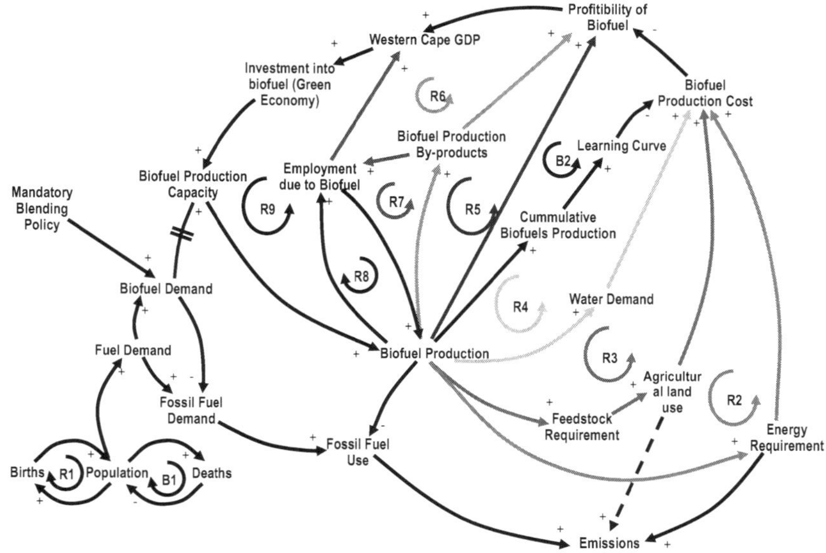

The CLD in Figure 5 shows the structure of the overall biofuel sector, by schematically showing how fuel demand is driven by population growth. The mandatory blending policy, acting as an exogenous driver, creates (+) a demand for biofuel, which in turn decreases (-) fossil fuel demand and fossil fuel use. A demand for biofuel is shown to lead to biofuel production capacity (+), which results in biofuel production. Biofuel production is one of the main drivers and increases (+) energy requirement, agricultural land use, water demand, by-product produced and employment – which ultimately leads back to an expansion of the biofuel production capacity (if biofuel production is financially lucrative). The CLD also indicates expected emissions being influenced by changes in fossil fuel use and the production of biofuel (through energy requirements and increased agricultural activities).

II. Dynamic modelling

The model-building phase resulted in the construction of eight core sub-models. These models are described in detail by Jonker[74] and include: (i)

73 Source: Jonker (2015).

biofuel production; (ii) agricultural yield; (iii) biofuel expenditure; (iv) operational finances; (v) profitability; (vi) alternative to local production; (vii) employment; and (viii) emissions. The eight sub-models are interconnected and represent the dynamics involved in biofuel production. As an example, the biofuel feedstock crop cost is extracted from the operational finances sub-model. The model shows that the biofuel feedstock crop cost will increase annually with the biofuel feedstock cost increase, which in turn is driven by a percentage increase (feedstock increase rate), as well as the effects of demand and supply on market prices. The time horizon of the simulation is 2001 to 2040, where the 2001 census data provided accurate initial values for the model.[75]

III. Scenario planning

Once the model was validated, a sensitivity analysis was done on variables that can be changed or determined endogenously. Electricity price was found to be one of the key endogenous variables influencing the success of local production (LP) scenarios. Different project scenarios were thus generated to look at the feasibility of generating electricity on-site and incorporating electricity sales as part of the business case. It was also necessary to look at the alternative solution for the Western Cape Province to form part of the mandatory blending requirement, which will see the province purchasing biofuel from existing plants elsewhere in the country, in other words relying on non-local production (NLP). For both local and non-local production, boiler-coal, boiler-biomass and combined heat and power (CHP) options were considered. The following six scenarios were compared to assess the impacts and effects of complying with the mandatory blending requirement on a provincial level:

1. LP (boiler-coal) – An LP scenario using a conventional coal-fired boiler for process heat
2. LP (boiler-biomass) – An LP scenario using a biomass-fired boiler for process heat
3. LP (CHP-coal) – An LP scenario where a combined heat and power (CHP) unit using coal to provide process heat and electricity (surplus can be sold to the national grid)

74 Jonker (2015).
75 (Ibid.).

4. LP (CHP-biomass) – Biomass is used instead of coal to fuel the CHP in scenario 3
5. NLP (coal) – An NLP scenario, which evaluates the effects of buying biofuel in from a different province, assuming that the said plant uses a coal-fired boiler
6. NLP (biomass) – An NLP scenario identical to scenario 5, but assuming the boiler is fuelled with biomass

IV. Modelling outcomes: Biofuel production

The biofuel production sub-model simulated bioethanol and biodiesel production capacity, as well as actual production volumes. The biofuel capacity model assumed that two facilities producing 160 million litres per annum of bioethanol and 35 million litres per annum of biodiesel will be completed by 2018.[76] The amount of cropland available for biofuel crop production was based on the inputs received from the agricultural sub-model in the Western Cape Green Economy Model (WeCaGEM). The biofuel production model assumed 25% of the current canola land to be allocated to biodiesel production and a four-year crop rotation cycle for triticale cultivation to produce bioethanol. The effects of farmer uptake of triticale cultivation was incorporated and based on the sale price of triticale compared to B3 or utility grade wheat. Experts suggest that an additional 70 000 ha of fallow land are available for triticale cultivation in the Western Cape Province, which contributed to the production capacity in the model. The model shows that actual production is less than production capacity and that production capacity will subsequently decrease, based on the reduction in required production capacity, primarily owing to limited biofuel feedstock available. The model also predicts that the biofuel production under the current feedstock assumptions is not sufficient to meet the biodiesel demand of the province based on the fuel demand and the mandatory blending requirements.

V. Modelling outcomes: Bottom-line costs

The bottom-line cost refers to the overall cost of complying with the blending mandate, and incorporates scenarios 5 and 6 (non-local produc-

76 GreenCape (2015).

tion), which will see the Western Cape Province buying biofuel from existing biofuel producers. The local and non-local production scenarios for bioethanol and biodiesel are shown in Figures 6A and 6B. For bioethanol (Figure 6A), it can be seen that the non-local production scenarios initially have lower associated costs, but will exceed the costs of all the local production scenarios by 2025, as a result of the continued expenditure on purchasing and transporting bioethanol with very little income being generated by the non-local production scenarios.

Figure 6: (A) Bottom-line cost for the production of (A) bioethanol and (B) biodiesel[77]

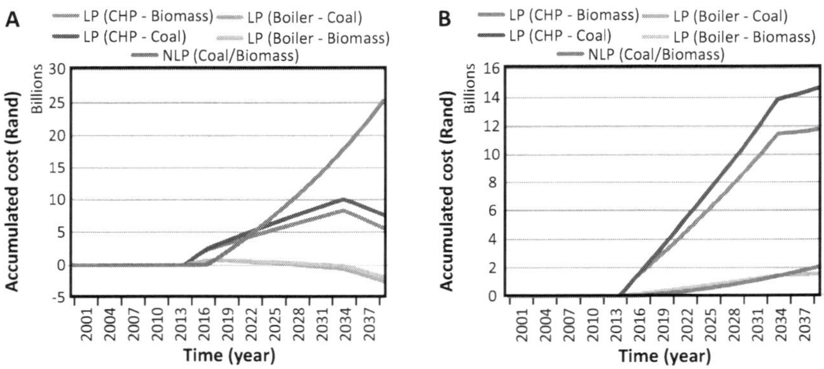

Figure 6B shows the bottom-line cost for biodiesel, where the cost of the local production scenarios making use of the CHP unit far exceeds the local production scenarios 1 and 2, as well as the non-local production scenarios. The non-local production bottom-line costs only marginally exceed scenarios 1 and 2 by 2036, meaning that for the majority of the estimated project lifespan it is expected to be cheaper to purchase biodiesel from existing production facilities, rather than producing biodiesel locally.

VI. Modelling outcomes: Employment

In evaluating employment creation through a biofuel sector, only direct additional job creation was considered. It is estimated that the bioethanol

77 Source: Jonker (2015).

plant will create roughly 2,080 jobs through the expansion of the agricultural sector and an average of 55 additional jobs related to operational duties. Biodiesel production will create relatively few jobs, namely an expected 10 to 15 additional jobs, owing to the small scale of the operation and the use of existing canola crops (thus not increasing agricultural activity).

VII. Modelling outcomes: Emissions

The emissions produced in various scenarios are mainly dependent on the process used and, as expected, the scenarios making use of energy from coal produced substantially more emissions. The bioethanol production facility utilising biomass produced 63% less emissions than the coal-fired counterpart. Biodiesel production using biomass showed a 5% reduction in emissions when compared to a coal-fired facility because of the small scale of the operation. For bioethanol and biodiesel the non-local production scenarios produced higher emissions than locally producing scenarios, owing to the additional emissions caused by road transportation of the biofuel from established facilities.

VIII. Scenarios summary and recommendations

All the scenarios were evaluated at the hand of the indicators identified (see Table 2). Recommendations to facilitate compliance with the mandatory blending policy in the Western Cape Province are summarised below and are discussed in detail in the work of Jonker.[78]

1. Bioethanol

The local production scenario 2 (boiler-biomass) is recommended, as an IRR of 23% is achieved, while emissions are 63% lower than the coal-fired option. To strengthen the business case of bioethanol production, the following factors should be addressed:

- Improve feedstock availability;

78 Jonker (2015).

- Reduce capital costs through alternative financing options; and
- Incorporate bioethanol production as part of an invasive alien land-clearing scheme, to supply process energy and avoid reliance on coal-based energy sources.

2. Biodiesel

It is advised that biodiesel not be locally produced (scenarios 5 and 6). The current operation scale and feedstock assumptions do not provide a compelling business case for local biodiesel production. Furthermore, the social and environmental advantages derived from biodiesel production do not provide sufficient incentive to justify the financial and infrastructure burdens associated with a project of this scale and nature. Further work is recommended in the following areas to establish more feasible biodiesel production options:

- Investigate different and combined feedstock options;
- Consider systems and policies to encourage small-scale biodiesel production; and
- Re-evaluate the value stream of biodiesel.

G. Conclusion and recommendations

Sub-Saharan Africa has vast resources of biomass available. The potential role of bioenergy/biofuels for a sustainable Africa should not be abandoned on the basis of established historical practices, a few examples of poor implementation, or the lack of political and financial support. Today, with an abundance of conversion technologies available, significant benefits can be achieved in all the sustainability categories through bioenergy production value chains – though this requires careful, quantitative assessments against indicators. Economic and legal frameworks must provide the appropriate tools and necessary checks and balances to ensure that preferred implementation approaches to bioenergy and biofuels are prioritised, while the less preferred approaches, having specific negative impacts, are excluded from future implementation. Examples of such economic models are provided for the Western Cape, but could be very useful for implementation in the broader sub-Saharan Africa context.

Bibliography

Ahlborg, H & L Hammar, 2014, Drivers and barriers to rural electrification in Tanzania and Mozambique – Grid-extension, off-grid, and renewable energy technologies, *Renew Energy* 61, 117.

Allen, JC & DF Barnes, 1985, The causes of deforestation in developing countries, *Ann Assoc Am Geogr* 75(2), 163.

Amigun, B, D Petrie & J Görgens, 2012, Feedstock and technology options for bioethanol production in South Africa: Techno economic prefeasibility study, *Energy & Fuels* 26(9), 5887.

Amigun, B, R Sigamoney & H von Blottnitz, 2008, Commercialisation of biofuel industry in Africa: A review, *Renew Sust Energ Rev* 12(3), 690.

Amigun, B & H von Blottnitz, 2011, Operating cost analysis of an annexed ethanol distillery in a rural, landlocked African country, *Environ. Prog Sustain Energy* 30(3), 500.

Anex RP, A Aden, FK Kazi, J Fortman, RM Swanson, MM Wright, JA Satrio, RC Brown, DE Daugaard, A Platon, G Kothandaraman, DD Hsu & A Dutta, 2010, Techno-economic comparison of biomass-to-transportation fuels via pyrolysis, gasification, and biochemical pathways, *Fuel* 89(1), S29.

Arvis JF, G Raballand & JF Marteau, 2007, *The cost of being landlocked: Logistics costs and supply chain reliability*, World Bank Report.

Batidzirai B, M Valk, B Wicke, M Junginger, V Daioglou, W Euler & ACP Faaij, 2016, Current and future technical, economic and environmental feasibility of maize and wheat residues supply for biomass energy application: Illustrated for South Africa, *Biomass Bioenerg* 92, 106.

Chidumayo, EN, DJ Gumbo, 2013, The environmental impacts of charcoal production in tropical ecosystems of the world: A synthesis, *Energy Sustain Dev* 17(2), 86.

Chimphango, AFA & JF Görgens, 2015, Postharvest technology for advancing sustainable bioenergy production for food processing, *Br Food J* 117(12), 2850.

Deenanathi, ED, S Iyuke, K Rumbold, 2012, The bioethanol industry in sub-Saharan Africa: History, challenges and prospects, *J Biomed Biotechnol* 2012, art. 416491.

DAFF / Department of Agriculture, Forestry and Fisheries, 2011, *A profile of the South African forestry and wood products market value chain*, at http://www.daff.gov.za/docs/AMCP/FOREWOODMVCP2011-12.pdf, accessed 25 July 2016.

Du Preez, J, 2016, *Assessment of process options for triticale fermentation to ethanol and animal feed in the Western Cape*, MEng thesis, Stellenbosch University.

Dwivedi P, JR Alavalapati & P Lal, 2009, Cellulosic ethanol production in the United States: Conversion technologies, current production status, economics, and emerging developments, *Energy Sustain Dev* 13(3), 174.

Eickhout, B, 2012, *A strategy for a bio-based economy*, Green New Deal Series Volume 9, Green European Foundation, at https://biobs.jrc.ec.europa.eu/sites/default/files/generated/files/policy/A%20Strategy%20for%20a%20Bio-Based%20Economy_Green%20MEPs_Green%20European%20Foundation.pdf}, accessed 31 July 2016.

Ejigu, M, 2008, Towards energy and livelihoods security in Africa: Smallholder production and processing of bioenergy as a strategy, *Nat Resour Forum* 32, 152.

El Fadel, M, G Rachid, R El-Samra, GB Boutros & Hashisho, 2013, Knowledge management mapping and gap analysis in renewable energy: Towards a sustainable framework in developing countries, *Renew Sust Energ Rev* 20, 576.

Fore, SR, P Porter, W Lazaru, 2011, Net energy balance of small-scale on-farm biodiesel production from canola and soybean, *Biomass Bioenerg* 35(5), 2234.

Fritsche, UR, K Hünecke, A Hermann, F Schulze, K Wiegmann & M Adolphe, 2006, *Sustainability standards for bioenergy*, WWF Germany, at http://www.biofuelstp.eu/downloads/WWF_Sustainable_Bioenergy_final_version.pdf, accessed 25 July 2016.

Gallagher, PW, WC Yee & HS Baumes, 2016, *2015 Energy balance for the corn-ethanol industry*, United States Department of Agriculture, Office of the Chief Economist, Office of Energy Policy and New Uses, February 2016.

GBEP / Global Bioenergy Partnership, 2011, *Sustainability indicators for bioenergy*, Food and Africulture Organisation (FOA) of the United Nations, at http://www.globalbioenergy.org/programmeofwork/task-force-on-sustainability/gbep-report-on-sustainability-indicators-for-bioenergy/en/, accessed 25 July 2016.

GCIS / The Government Communication and Information System, 2015, *South Africa yearbook 2014/15: Agriculture*, at http://www.gcis.gov.za/sites/www.gcis.gov.za/files/docs/resourcecentre/Agriculture2015.pdf], accessed 24 Jul 2016.

Görgens, JF, M Carrier, MP García-Aparicio, 2014, "Biomass conversion to bioenergy products", in: T Seifert (ed), *Bioenergy from wood, managing forest ecosystems*, volume 26, 2014, 137.

GreenCape, 2013, *Developing a business case for sustainable biofuels in South Africa: A focus on waste-based bioethanol for fleet transport in the Western Cape*, at http://www.grainsa.co.za/upload/Final-Report-Bioethanol-in-the-Western-Cape-November-2013.pdf, accessed 25 July 2016.

GreenCape, 2015, *Biofuels: from viability to pilot project (re-assessment of the Western Capes biofuel production potential via a multi-criteria analysis)*, at http://greencape.co.za/what-we-do/projects/biofuels/, accessed 14 January 2016.

Haanyika, CM, 2008, Rural electrification in Zambia: A policy and institutional analysis, *Energy Policy* 36(3), 1044.

Hassan, SF, 2008, Development of sugar industry in Africa. *Sugar Tech* 10, 197–203.

Ho, DP, HH Ngo & W Guo, 2014, A mini review on renewable sources for biofuel, *Bioresour Technol* 169, 742.

Hugo, W, 2015, *BioEnergy feasibility in South Africa,* EGU General Assembly 2015, held 12-17 April, 2015 in Vienna, Austria. Published April 2015, at http://www.saeon.ac.za/news-updates-archives/BioEnergy%20Atlas%20for%20Paris2015.pdf, accessed 25 July 2016.

IEA / International Energy Agency, 2014, *Africa energy outlook: A focus on energy prospects in sub-Saharan Africa*, at https://www.iea.org/publications/freepublications/publication/WEO2014_AfricaEnergyOutlook.pdf, accessed 24 July 2016.

IEA-RETD / The International Energy Agency's Implementing Agreement for Renewable Energy Technology Deployment, 2016, *Towards advanced biofuels - options for integrating conventional and advanced biofuel production sites (RES-T-BIOPLANT)*, [Ugarte, S, U Fritsche, SQ Consult B.V. and IINAS GmbH], IEA Implementing Agreement for Renewable Energy Technology Deployment (IEA-RETD), at http://iea-retd.org/wp-content/uploads/2016/08/20160202-IEA-RETD-RES-T-BIOPLANT-1.pdf, accessed 30 November 2016.

IRENA / International Renewable Energy Agency, 2015, *Africa 2030: Roadmap for a renewable energy future*, at http://www.irena.org/DocumentDownloads/Publications/IRENA_Africa_2030_REmap_2015_low-res.pdf, accessed 24 July 2016.

Jonker, WD, 2015, *Biofuel implications of a green economy transition in the Western Cape province of South Africa: A system dynamics modelling approach to biofuel*, Master thesis, Stellenbosch University.

Jonker, WD, AC Brent & JK Musango, 2015, *Modelling the production of biofuel within the Western Cape, South Africa*, International Association for Management of Technology (IAMOT).

Krause, K, 2010, *The true size of Africa*, at http://kai.sub.blue/en/africa.html, accessed 24 July 2016.

Leibbrandt, NH, JH Knoetze & JF Görgens, 2011, Comparing biological and thermochemical processing of sugarcane bagasse: An energy balance perspective, *Biomass Bioenerg* 35(5), 2117.

Lucas, PL, J Nielsen, K Calvin, DL McCollum, G Marangoni, J Strefler, BCC van der Zwaan & DP van Vuuren, 2015, Future energy system challenges for Africa: Insights from integrated assessment models, *Energy Policy* 86, 705.

Lynd, LR, M Sow, AFA Chimphango, LAB Cortez, CHB Cruz, M Elmissiry, M Laser, IA Mayaki, MAFD Moraes, LAH Nogueira, GM Wolfaardt, J Woods & WH van Zyl, 2015, Bioenergy and african transformation, *Biotechnol Biofuels* 8, 18.

Lynd, LR, H von Blottnitz, B Tait, J De Boer, IS Pretorius, K Rumbold & WH van Zyl 2003, Plant biomass conversion to fuels and commodity chemicals in South Africa: A third hhapter? *S Afr J Sci* 99, 499.

LR Lynd & J Woods, 2011, A new hope for Africa, *Nature* 474, S20.

Mafakheri F & F Nasiri, 2014, Modeling of biomass-to-energy supply chain operations: Applications, challenges and research directions, *Energy Policy* 67, 116.

Meincken, M & L Tyhoda, 2014, "Biomass quality" in: T Seifert (ed), *Bioenergy from wood, managing forest ecosystems*, Volume 26, 169.

Mirzabaev, A, D Guta, J Goedecke, V Gaur, J Börner, D Virchow, M Denich & J von Braun, 2014, *Bioenergy, food security and poverty reduction: Mitigating tradeoffs and promoting synergies along the water-energy-food security nexus.* ZEF Working Paper Series, Department of Political and Cultural Change, Center for Development Research, University of Bonn.

Mugido, W, J Blignaut, M Joubert, J De Wet, A Knipe, S Joubert, B Cobbing, J Jansen, D Le Maitre & M van der Vyfer, 2014, Determining the feasibility of harvesting invasive alien plant species for energy, *S Afr J Sci* 110, 11.

Munyinda, K, FD Yamba & H Walimwipi, 2011, "Chapter 8: Bioethanol potential and production in Africa: Sweet sorghum as a complementary feedstock", in: R Janssen & D Rutz (eds), *Bioenergy for sustainable development in Africa*, 81.

Musango, JK, AC Brent, JBS Van Niekerk, WD Jonker, AP Pienaar, TA York, J Oosthuizen, L Duminy & IH de Kock, 2015, *A system dynamics approach to understand the implications of a green economy transition in the Western Cape Province of South Africa.* 33rd International Conference of the System Dynamics Society, Boston, USA.

Nolte, M, 2007, *Commercial biodiesel production in South Africa: A preliminary economic feasibility study*, PhD thesis, Stellenbosch University.

Padi, RK, 2016, *Potential and economic impact of renewable energy in improving African rural food processing*, MEng Thesis, Stellenbosch University.

Pan, SY, MA Du, IT Huang, IH Liu, EE Chang & PC Chiang, 2015, Strategies on implementation of waste-to-energy (WTE) supply chain for circular economy system: A review, *J Clean Prod* 108 (Part A), 409.

Panwar, NL, R Kothari & VV Tyagi, 2012, Thermochemical conversion of biomass – Eco friendly energy routes, *Renew Sust Energ Rev* 16(4), 1801.

Sarkar, N, SK Ghosh, S Bannerjee & K Aikat, 2012, Bioethanol production from agricultural wastes: An overview, *Renew Energy* 37(1), 19.

Sarrafa, M, B Rismanchi, R Saidur, HW Ping & NA Rahim, 2013, Renewable energy policies for sustainable development in Cambodia, *Renew Sust Energ Rev* 22, 223.

SAPIA / South African Petroleum Industry Association, 2014, *2014 Annual Report* at http://www.sapia.org.za/Portals/0/Annual-Reports/SAPIA_AR_2014.pdf?ver=2015-12-08-112603-243, accessed 24 July 2016.

Sawatdeenarunat, C, KC Surendra, D Takara, H Oechsner & SK Khanal, 2015, Anaerobic digestion of lignocellulosic biomass: Challenges and opportunities, *Bioresour Technol* 178, 178.

Schut, M & MJ Florin, 2015, The policy and practice of sustainable biofuels: Between global frameworks and local heterogeneity. The case of food security in Mozambique. *Biomass Bioenerg* 72, 123.

Schut, M, NC Soares, G van de Ven & M Slingerland, 2014, Multi-actor governance of sustainable biofuels in developing countries: The case of Mozambique, *Energy Policy* 65, 631.

Smith, JU, A Fischer, PD Hallett, HY Homans, P Smith, Y Abdul-Salam, HH Emmerling, & E Phimister, 2015, Sustainable use of organic resources for bioenergy, food and water provision in rural Sub-Saharan Africa, *Renew Sust Energ Rev* 50, 903.

Stafford, WHL & AC Brent, 2011, Bioenergy systems sustainability assessment and management, *Renew Energy Law & Policy* 2(3), 205.

Subedi, M, RB Matthews, M Pogson, A Abegaz, BB Balana, J Oyesiku-Blakemore & J Smith, 2014, Can biogas digesters help to reduce deforestation in Africa? *Biomass Bioenerg* 70, 87.

UN / United Nations, 2012, *The millennium development goals report 2012*, at http://www.un.org/millenniumgoals/pdf/MDG%20Report%202012.pdf, accessed 24 July 2016.

UN-Habitat / United National Human Settlements Programme, 2014, *State of African cities 2014, re-imagining sustainable urban transitions*, at http://unhabitat.org/books/state-of-african-cities-2014-re-imagining-sustainable-urban-transitions/, accessed 25 July 2016.

Van Zyl, WH, AFA Chimpangho, R Den Haan, JF Görgens & PWC Chirwa, 2011a, Next generation cellulosic ethanol technologies and their contribution to a sustainable Africa, *Interface Focus* 1, 196.

Van Zyl, WH, R Den Haan & DC La Grange, 2011b, Developing organisms for consolidated bioprocessing of biomass to ethanol, in: MA dos Santos Bernardes (ed), *Biofuels Production – Recent Developments and Prospects,* 137, at http://www.intechweb.org/, accessed 24 July 2016.

Von Maltitz, G, L Haywood, M Mapako & AC Brent, 2009, *Analysis of opportunities for biofuel production in sub-Saharan Africa*, CIFOR Environment Brief, at http://www.cifor.cgiar.org/publications/briefs, accessed 25 July 2016.

Watson, HK, 2011, Potential to expand sustainable bioenergy from sugarcane in southern Africa, *Energy Policy* 39, 5746.

Winkler, H, AF Simões, EL la Rovere, M Alam, A Rahman & S Mwakasonda, 2011, Access and affordability of electricity in developing countries, *World Dev* 39(6), 1037.

Zhang, C, G Xiao, L Peng, H Su & T Tan, 2013, The anaerobic co-digestion of food waste and cattle manure, *Bioresour Technol* 129, 170.

Biofuel plants as potential invasive species: Environmental concerns and progress towards objective risk assessment

Dr Ryan Blanchard, Natural Resources and the Environment, CSIR, Stellenbosch, South Africa, and Centre for Invasion Biology, Department of Botany and Zoology, Stellenbosch University, South Africa

Dr Sabrina Kumschick, Centre for Invasion Biology, Department of Botany and Zoology, Stellenbosch University, South Africa, and Invasive Species Programme, South African National Biodiversity Institute, Kirstenbosch Research Centre, Cape Town, South Africa

Prof Dr David M Richardson, Centre for Invasion Biology, Department of Botany and Zoology, Stellenbosch University, South Africa

A. Introduction

The uncertainty of long-term fossil fuel supply and volatile fuel prices, along with increasing CO_2 emissions, has generated interest in alternative fuel sources, especially biofuels. Biofuel production is being widely advocated as a renewable and environmentally friendly way of reducing the use of fossil fuels. An important driver of biofuel production in many developing countries – besides that of potentially reducing greenhouse gas emissions and enhancing fuel security – is the need to facilitate development, provide jobs, and improve trade balances. There are, however, many environmental concerns relating to the rapid increase in biofuel production. Accommodating the level of biofuel production that would be required to achieve a substantial reduction in the use of fossil fuels, among other aims, will impose a major strain on the sustainable use of different land types and will place the agricultural and conservation sectors under major pressure. Besides the threat to the environment through radical increases in habitat conversion, another threat associated with biofuel production is the invasive spread of feedstock species beyond areas set aside for biofuel production. Most biofuel feedstock species are not indigenous to the areas where they are likely to be grown, and many of the most suitable feedstock species are either known to be invasive elsewhere in the world or have features that are strongly associated with invasiveness.

Biological invasions – i.e. the establishment and spread of alien species in areas to which they have been introduced through human action – are a large and growing threat to biodiversity and ecosystem functioning worldwide. Species from all taxonomic groups have been moved around the world and thousands of species have negative impacts on ecosystems where they spread, proliferate and disrupt ecosystem functioning in many ways.[1] The numbers of invasive species, the extent of invasions, and the types and overall magnitude of impacts caused by invasive species are increasing rapidly. Many species that are currently invasive were introduced accidentally, but an increasingly large proportion of invaders, including many of the most damaging species, was intentionally introduced. Most of these were spread around while providing a variety of benefits to humans. Some of these species are still both useful and harmful in the same region, usually bringing benefits to a small sector of society, but causing harmful and sometimes irreversible effects whose costs are generally borne by society at large. Aquaculture, forestry, horticulture and the pet trade are examples of sectors that have 'sown the seeds' of major invasions in many parts of the world. These sectors have all taken steps to reduce problems associated with the species on which their enterprises depend. The emerging biofuel enterprise is a 'new kid on the block' in this regard, and measures to improve the sustainability of the endeavour by reducing problems with invasive feedstock species are only beginning to be debated and implemented. This chapter looks at progress in invasion science and how this can be applied to reduce problems associated with invasions emanating from biofuel plantings. Special attention is given to the situation in South Africa.

B. *Current progress towards predicting invasiveness and invasibility*

Differentiating between alien plant species that have a high risk of becoming invasive and those with a low risk of invasiveness is a significant challenge, and the management of potential impacts of invasive species demands close collaboration between ecologists, economists and policy makers.[2] Many factors determine whether an introduced species will become invasive in a new environment: these include species-specific traits, environmental factors, a diverse array of human-mediated factors, and

1 Pyšek & Richardson (2010).
2 Anderson et al. (2004).

complex interactions between these drivers. Unfortunately there are no single species' traits or syndrome of traits that consistently predict invasiveness for all plant species.[3] Nonetheless, much progress has been made towards identifying traits associated with plant invasiveness[4] and using such insights to develop approaches for predicting invasiveness.[5] Biological traits that are common in most invasive plant species, many of which characterise some biofuel crops,[6] include: fast growth rates, rapid reproduction (e.g. high seed output), short minimum generation times, and a tolerance of a wide range of environmental habitats. Other traits that are general indicators of plant invasiveness include smaller seeds, vegetative growth, self-fertilisation, and a high capacity for dispersal (e.g. by animals or wind). The importance of all these traits varies considerably among different plant groups. Features of the receiving environment also play a crucial role in determining whether an invasion will occur.[7] A habitat is invasible when an alien species is able to establish and sustain its current range on the site of introduction or expand its range beyond the initial site.[8] The intentional introducers of species (e.g. for forestry or biofuel production) often use climatic and other data to identify suitable locations in which particular species are likely to flourish – and such environments are likely to be well suited for the species to reproduce and spread from introduction sites. Degraded or highly disturbed areas are generally more likely to be invaded than are intact areas. A crucial factor influencing the likelihood of a species becoming invasive is propagule pressure – the number of propagules (usually seeds) that are available to drive population growth and spread. High levels of propagule pressure greatly increase the likelihood of an alien species becoming established (developing self-sustaining populations) and invasive (spreading from cultivation sites). A successful invasion is the result of a multiple-step process whereby numerous barriers must be overcome prior to a species becoming invasive.[9] Different pathways of introduction of alien species create different barriers, but also particular windows of opportunity for alien species to become established and spread. For example, plant species used in forestry and for biofuel production are usually planted in very large numbers (which assure high propa-

3 Kolar & Lodge (2001).
4 Baker (1965); Alpert (2006); Pyšek & Richardson (2007).
5 Pheloung et al. (1999); Pattison & Mack (2008).
6 Raghu et al. (2006).
7 Richardson & Pyšek (2006); Lewis & Porter (2014).
8 Burke & Grime (1996).
9 Blackburn et al. (2011).

gule pressure) and in configurations (plantations) that provide superb "launching pads" for invasions.[10]

C. Biofuel crops as weeds

Numerous studies show that most plant species under consideration for widespread planting to produce biofuels pose a high risk of becoming invasive.[11] While some species are already well-known invaders, others have traits that are likely to allow them to become invasive if introduced to new regions and cultivated in large numbers. New industrial requirements demand that feedstocks targeted for cellulose production are fast growing, easy to establish, and require few additional inputs – the same traits that are associated with invasiveness. The concern about introducing potential invasive species to new locations around the world has prompted widespread discussion on how biofuels could add to the invasive species problem.[12] Much research has been done on developing tools and guidelines to reduce future problems with invasive biofuel plants.[13]

The magnitude of the problem caused by the cultivation of alien plants in similar industries such as forestry and agroforestry over previous decades[14] has led to the development of approaches to manage problems of invasiveness of biofuel plants. The success of these protocols can be measured by the uptake of the legislative requirements, which apply to the introduction of new species. Lessons on how these rules are influencing the biofuels industry are beginning to emerge. The use of protocols such as risk assessments in legislation is becoming standard practice in many countries.

D. The role of legislation in managing invasions

Owing to the large number of plant species being moved and traded around the world, international conventions have been established to pro-

---

10 Richardson (2011); Richardson & Blanchard (2011).
11 E.g. Low & Booth (2007); Buddenhagen et al. (2009).
12 E.g. Low & Booth (2007).
13 E.g. Lonsdale & FitzGibbon (2011); Richardson & Blanchard (2011); Barney (2014); Barney et al. (2015)
14 Richardson (2011).

tect countries from introducing or exporting potentially harmful species.[15] These agreements provide an incentive for countries to administer national laws. They are enforced to govern the introduction and spread of invasive species and to minimise their impacts on economic activities such as agriculture and local biodiversity. The International Plant Protection Convention (IPPC), the World Trade Organization's Agreement on Sanitary and Phytosanitary Measures and the Convention on Biological Diversity (CBD) are examples of important conventions that regulate the movement, trade and potential impacts of alien and invasive species.[16] Biofuels have triggered responses from international organisations such as the International Union for the Conservation of Nature (IUCN), the United Nations Environment Programme (UNEP) and the Centre for Agricultural Biosciences International (CABI) to further implement regulations to address the problem of introducing invasive alien species.

In South Africa these conventions and resulting national legislation that has been enacted over many years have been interpreted to provide decisions on species introductions that avoid negative economic, social and environmental impacts. However, the promulgated laws and acts have been fragmented and spread across different departments, authorities and institutions – and this hampers the express goal of preventing and managing biological invasions. The adoption of the Alien and Invasive Species Regulations of the National Environmental Management: Biodiversity Act in 2014 aims to provide an integrated approach to restricting, managing and controlling alien and invasive species.

An important feature of the South African regulations concerning alien and invasive species is the listing of plant species into different categories based on current impacts, abundance, ease of control and commercial use. All applications for new species introductions should be subjected to an assessment to determine whether listing is necessary (See *Risk assessment* below). The categories are:

- Category 1(a): Alien and invasive species that are targeted for nationwide eradication. This category includes alien species where effective control and eradication is possible.
- Category 1(b): Alien and invasive species that require on-going control as part of a management plan. The spread of alien species in this

15 Richardson et al. (2003)
16 (Ibid.).

category must be contained or efforts must be made to bring them under control.

- Category 2: Alien and invasive species that require a permit and can only be used under restricted conditions. Their spread needs to be contained, and all individual plants without a permit need to be controlled. This category includes alien species that have commercial or other perceived value.
- Category 3: Alien and invasive species that may be retained without a permit, but where propagation and trade are prohibited, and where spread must be contained. This category includes alien species that may remain in prescribed areas, but planting, propagation and trade is prohibited.
- Prohibited: This category includes alien species that do not yet occur in South Africa, and for which a permit for import will not be granted. If a prohibited species were to be found in the country, it would automatically become a Category 1(a) listed species.

Apart from the legal instruments listed above which apply to biofuels, various international initiatives seek to enhance the sustainability and to improve the public perception of the biofuels sector. For example the Roundtable on Sustainable Biomaterials (RSB)[17] recognises the potential impact of plant species used for biofuel production to become invasive species and aims to provide guidelines across the whole production cycle for producers and processors. These are voluntary guidelines that seek to ensure compliance with local and international regulations.

E. Assessing and managing risks associated with invasiveness of biofuel plants

The national and international agreements and regulations mentioned above require risk assessments to be performed to restrict trade and for new introductions.

Risk assessments are tools developed to prevent future damage caused by specific hazards. They address the likelihood of a hazard (in our case, invasiveness of alien biofuel species) occurring, and the consequences (i.e. negative impacts) thereof. Besides the legal requirement to perform formal risk assessments for new introductions, these assessments have also been

---

17 See www.rsb.org for further information.

shown to be economically beneficial as they can prevent costs arising from damage caused by alien species. Performing risk assessments and taking preventative action is cheaper than having to deal with the unwanted consequences of an alien species becoming a damaging invader.[18]

Many risk assessment schemes have been developed specifically to assess the risks associated with alien species.[19] Some of these schemes are also applicable for the introduction and production of biofuels. The most prominent scheme, which has been used to assess risks of biofuels in many parts of the world, is the so-called Australian Weed Risk Assessment.[20] It uses information on various biological and ecological traits of the species under consideration and information on its behaviour elsewhere. Answering a number of questions results in a score, which denotes the risk of the species becoming invasive. Because plants suitable for biofuel production need to be fast-growing, highly productive, and well suited to the climatic conditions where they are grown, they usually get high scores in such assessments, which results in their being flagged as high-risk introductions.

It is important to note that risk assessment schemes generally focus only on potential negative effects, as they are specifically developed to highlight problems which could arise from importing and planting a species. Many alien species have obvious benefits for forestry, horticulture, and as biofuels, to give a few examples. This means that the results of a risk assessment such as performed with the Australian Weed Risk Assessment need not result in the outright ban of high-risk species. High-risk species could still be used, under certain conditions and following the implementation of appropriate measures to alleviate risks. To achieve a trade-off between benefits and potential harm caused by the species, management options to reduce harm while retaining benefits need to be assessed and implemented.[21]

Risk management is the evaluation and selection of options to reduce the risk of an alien species. Such considerations include the successful management options available for a specific species. For biofuel production that can include proactive measures to prevent spread, including an early investment in biological control and genetic engineering where appropriate, and planning the design and configuration of plantations to min-

18 E.g. Keller et al. (2007).

19 Leung et al. (2012); Kumschick & Richardson (2013).

20 Pheloung et al. (1999); Buddenhagen et al. (2009); Gordon et al. (2011) and (2016).

21 E.g. Cousens (2008).

imise the spread of risk (e.g. by taking account of dispersal corridors and sensitive habitats). Mitigation plans also provide for regular monitoring of establishment and spread, and regular revision of management plans to incorporate international best practice and local knowledge.[22] It is crucial to assess the entire production chain to discover where the risks are greatest, to determine the possible consequences and identify the options needed to manage these. As mentioned above, South African legislation specifically allows for such risky but beneficial species to be imported and then to be used under specific conditions, as outlined in a permit which may, for example, stipulate the need to control all individual plants escaping from designated plantations, or require plants to be harvested before they produce seeds.

The entire process of risk assessment and risk management is termed *risk analysis*, and this typically also includes *risk communication* (Figure 1). Communicating and interacting with stakeholders is crucial throughout the whole risk analysis process to achieve understanding of the risk assessment conclusions and decisions made. Furthermore, input from stakeholders is crucial to reach a viable decision on how to manage certain species. Insights from engagement with stakeholders in other sectors, such as the ornamental horticulture industry,[23] will be useful in this regard. It is important to realise that different stakeholders with differing value systems will be affected differently by the threat and encroachment of alien species, as well as by the decisions taken in terms of their management. Obtaining support for these decisions is therefore crucial to ensure successful outcomes.

---

22 Richardson & Blanchard (2011).
23 E.g. Novoa et al. (2016).

Figure 1: The risk analysis process

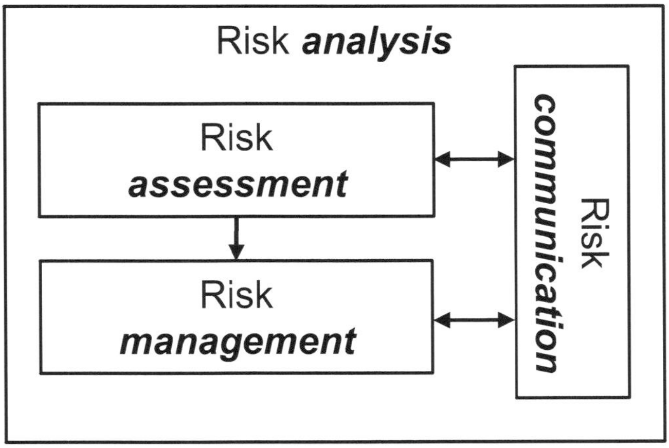

F. *Recommended policy approaches*

This section will elucidate some important requirements to minimise the spread of alien species. Experiences are drawn from South Africa, a country with a long history of dealing with alien and invasive species, including many species that are both commercially important crops and damaging invasive species.[24] Unfortunately, South Africa is the only country in sub-Saharan Africa that manages to fund the active clearing of invasive alien plants and is one of a few countries to have instituted a system for intercepting the arrival of invasive species.[25]

There are multiple potential pathways for species to enter a country. It is assumed that the selection of species for use as biofuel feedstocks will follow one of two pathways: 1) the deliberate introduction of a potential biofuel crop; or 2) the utilisation of a species (native or alien) already present in the country. The best way of reducing impacts of invasive species is to prevent their introduction in the first place. 'Pre-border' risk assessment protocols (RAPs) are designed to assess applications that require the introduction of species not currently in the country for commercial purposes. RAPs draw on global lists of known invaders and other information to assess the risk of a species becoming invasive (see above). The extensive literature on plant invasions and the increasing availability of online

24 E.g. van Wilgen et al. (2011); van Wilgen & Richardson (2012).
25 Boy & Witt (2013).

databases of invasive species means that it is, in most cases, easy to determine whether a potential biofuel species is invasive elsewhere. Such information and other provisions of RAPs need to be adapted to fit the particular requirements for risk assessment in the national legislation of different countries. Where species that are already in a country are selected for use in biofuel production, environmental impact assessments (EIAs) will normally be required as part of general environmental legislation. Attention must be given to ensuring that such EIAs apply international best practice tools to identify and mitigate risks associated with invasiveness. Numerous authors agree that a multi-tiered approach is needed for such risk analysis.[26] Linking such protocols with existing legislative requirements in different countries is an urgent priority.

There is usually a substantial time lag between the introduction and dissemination of an alien species and the manifestation of invasions. This has important implications in respect of liability for damages that may result owing to invasions of species intentionally introduced and cultivated for a specific purpose. Legislation needs to make provision for liability to be placed on the importers or the applicant of a permit for such introductions. Provisions for post-entry monitoring of approved biofuel species in buffer areas adjacent to projects may indicate that a species is capable of spreading and should result in stricter management protocols. This would ensure that due diligence and the costs of minimising environmental risks are captured in the overall assessment of the viability of biofuel projects.

G. Conclusion

For parts of the world where biofuel enterprises have not yet been established, the process of feedstock selection must be informed by adequate risk analysis. Hard lessons learnt in fields such as forestry and agroforestry, whose economic model requires the cultivation of alien plants in new environments, can be applied to avoid some of the pitfalls experienced. Screening protocols and global databases of well-known biofuel crops are beginning to emerge; these are providing the means for more informed and objective decisions to be made regarding risks of invasiveness and the impacts associated with species selected for biofuel production.

Key lessons from invasion science emphasise that the biological characteristics of a species (e.g. reproductive capability, species invasion his-

26 Quinn et al. (2013); Gordon et al. (2016).

tory and environmental tolerance) have predictive value. However, invasions are driven by many other factors, all of which need to be considered in objective assessments.

The introduction of economically important new crops, such as biofuel feedstocks, has led to the recent advances in the development of broader risk analysis protocols. While weed risk assessments are a critical screening component, the ability to embed such assessments within a decision support system is crucial. In other words, it is necessary to focus on the risks, benefits and the options available to manage risks across the entire biofuel value chain (e.g. transport, establishment and processing) in order to improve sustainability. In the view of the authors of this chapter, objectivity and sustainability can be achieved when the policies and practices linked to biofuel production aim to balance economic growth, while protecting natural ecosystems.

Bibliography

Alpert, P, 2006, The advantages and disadvantages of being introduced, *Biological Invasions* 8, 1523.

Anderson, MC, H Adams, B Hope & M Powell, 2004, Risk assessment for invasive species, *Risk Analysis* 24, 787.

Baker, HG, 1965, "Characteristics and modes of origin of weeds", in: HG Baker & GL Stebbins (eds), *The genetics of colonizing species*, 147.

Barney, JN, 2014, Bioenergy and invasive plants: Quantifying and mitigating future risks, *Invasive Plant Science and Management* 7, 199.

Barney, JN, LL Smith & DR Tekiela, 2015, "Using weed risk assessments to separate the crops from the weeds", in: LD Quinn, DP Matlaga & JN Barney (eds), *Bioenergy and biological invasions: Ecological, agronomic and policy perspectives on minimizing risk*, 67.

Blackburn, TM, P Pyšek, S Bacher, JT Carlton, RP Duncan, V Jarošík, JRU Wilson & DM Richardson, 2011, A proposed unified framework for biological invasions, *Trends in Ecology & Evolution* 26, 333.

Boy, G & A Witt, 2013, *Invasive alien plants and their management in Africa*, UNEP/GEF Removing barriers to invasive plant management project, International Coordination Unit CABI Africa, at https://portals.iucn.org/library/efiles/documents/2013-035.pdf, accessed 30 November 2016.

Buddenhagen, CE, C Chimera & P Clifford, 2009, Assessing biofuel crop invasiveness: A case study, *PLoS One* 4(4), e5261.

Burke, MJW, JP Grime, 1996, An experimental study of plant community invisibility, *Ecology* 77, 776.

Cousens, R, 2008, Risk assessment of potential biofuel species: An application for trait-based models for predicting weediness? *Weed Science* 56, 873.

Gordon, DR, KJ Tancig, DA Onderdonk & CA Gantz, 2011, Assessing the invasive potential of biofuel species proposed for Florida and the United States using the Australian weed risk assessment, *Biomass and Bioenergy* 35, 74.

Gordon, DR, SL Flory, D Lieurance, PE Hulme, C Buddenhagen, B Canton, PE Champion, TM Culley, C Daehler, F Essl, JE Hill, RP Keller, L Kohl, AL Koop, S Kumschick, DM Lodge, RN Mack, LA Meyerson, GR Pallipparambil, FD Panetta, P Pyšek, LD Quinn, DM Richardson, D Simberloff & M Vilà, 2016, Weed risk assessments are an effective component of invasion risk management, *Invasive Plant Science Management* 9, 80.

Keller, RP, DM Lodge & DC Finnoff, 2007, Risk assessment for invasive species produces net bioeconomic benefits, *Proceedings of the National Academy of Sciences of the USA* 104, 203.

Kolar, CS & DM Lodge, 2001, Progress in invasion biology: predicting invaders, *Trends in Ecology & Evolution* 16, 199.

Kumschick, S & DM Richardson, 2013, Species based risk assessments for biological invasions: advances and challenges, *Diversity and Distributions* 19, 1095.

Leung, B, N Roura-Pascual, S Bacher, J Heikkilä, L Brotons, MA Burgman, K Dehnen-Schmutz, E Essl, PE Hulme, DM Richardson, D Sol & M Vilá, 2012, TEASIng apart alien-species risk assessments: a framework for best practices, *Ecology Letters* 15, 1475.

Lewis, KC & RD Porter, 2014, Global approaches to addressing biofuel-related invasive species risks and incorporation in US laws and policies, *Ecological Monographs* 84, 171.

Lonsdale, WM & F FitzGibbon, 2011, The known unknowns – managing the invasion risk from biofuels, *Current Opinion in Environmental Sustainability* 3, 31.

Low, T & C Booth, 2007, *The weedy truth about biofuels*, The Invasive Species Council.

Novoa, A, H Kaplan, JRU Wilson & DM Richardson, 2016, Resolving a prickly situation: Involving stakeholders in invasive cactus management in South Africa, *Environmental Management* 57, 998.

Pheloung, PC, PA Williams & SR Halloy, 1999, A weed risk assessment model for use as a biosecurity tool evaluating plant introductions, *Journal of Environmental Management* 5, 239.

Pattison, RR & RN Mack, 2008, Potential distribution of the invasive tree *Triadica sebifera* (Euphorbiaceae) in the United States: Evaluating CLIMEX predictions with field trials, *Global Change Biology* 14, 813.

Pyšek, P & DM Richardson, 2007, Traits associated with invasiveness in alien plants: Where do we stand?, in: W Nentwig (ed), *Biological Invasions*, Ecological Studies 193, 97.

Pyšek, P & DM Richardson, 2010, Invasive species, environmental change and management, and ecosystem health, *Annual Review of Environment and Resources* 35, 25.

Quinn, L, JN Barney, J McCubbins & A Endres, 2013, Navigating the "noxious" and "invasive" regulatory landscape: Suggestions for improved regulation, *BioScience* 63, 124.

Raghu, S, RC Anderson, CC Daehler, AS Davis, RN Wiedenmann, D Simberloff & RN Mack, 2006, Adding biofuels to the invasive species fire? *Science* 313, 1742.

Richardson, DM, 2011, "Forestry and agroforestry", in: D Simberloff & M Rejmánek (eds) *Encyclopedia of biological invasions*, 241.

Richardson, DM & R Blanchard, 2011, Learning from our mistakes: minimizing problems with invasive biofuel plants, *Current Opinion in Environmental Sustainability* 3, 36.

Richardson, DM, JA Cambray, RA Chapman, WRJ Dean, CL Griffiths, DJ Newton & TJ Winstanley, 2003, "Vectors and pathways of biological invasions in South Africa: Past, present and future", in: G Ruiz & JT Carlton (eds), *Invasive Species Vectors and Management Strategies*, 292.

Richardson, DM & P Pyšek, 2006, Plant invasions: merging the concepts of species invasiveness and community invisibility, *Progress in Physical Geography* 30, 409.

Van Wilgen, BW, C Dyer, JH Hoffmann, P Ivey, DC Le Maitre, JL Moore, DM Richardson, M Rouget, A Wannenburgh & JRU Wilson, 2011, National-scale strategic approaches for managing introduced plants: insights from Australian acacias in South Africa, *Diversity and Distributions* 17, 1060.

Van Wilgen, DM & BW Richardson, 2012, Three centuries of managing introduced conifers in South Africa: Benefits, impacts, changing perceptions and conflict resolution. *Journal of Environmental Management* 106, 56.

PART II: BIOFUELS – SCIENCE FOR SOCIETY, POLICY FOR DEVELOPMENT

The nexus between energy security, biofuels and climate change: Insights from the Fifth Assessment Report of the Intergovernmental Panel on Climate Change

Prof Dr Oliver C Ruppel, Director of the Climate Change Policy and Energy Security Programme for Sub-Saharan Africa of the Konrad-Adenauer-Foundation; Professor of Law, Development and Rule of Law Programme (DROP), Faculty of Law, University of Stellenbosch, South Africa.

A. Introduction

Recent literature shows increasing attention to questions of ethics, justice, and responsibilities relating to climate change and energy security. As basic resources such as energy, land, food, and water become more and more threatened, inequalities and unfairness may deepen. The Intergovernmental Panel on Climate Change (IPCC) Special Report on Renewable Energy[1] has already highlighted the uncertainties around the role of biofuels in food price increases and risks of deteriorating food security with future deployment of bioenergy. Increasing demand for biofuels shifts land from food to fuel production, which may increase food prices and affect the poor negatively. On the other hand, there are many compelling reasons for countries to encourage biofuel production. Biofuels, for instance, are a means to exert downward pressure on oil prices, to stimulate rural development and to reduce oil. No doubt, more research is needed to understand fully the socioeconomic and environmental tradeoffs associated with biofuel production in Africa. Meanwhile, the 2014 Fifth Assessment Report on Climate Change (AR5) of the IPCC is most probably one of the most policy-relevant documents currently at hand.

1 IPCC (2012).

The following sections give a brief introduction to the work of the IPCC and AR5's coverage of energy security, bioenergy and biofuels, with a particular focus on southern Africa.[2]

B. The Intergovernmental Panel on Climate Change

The Intergovernmental Panel on Climate Change (IPCC) was established by the United Nations Environment Programme (UNEP) and the World Meteorological Organisation (WMO) in 1988. The ultimate role of the IPCC is –[3]

> to assess on a comprehensive, objective, open and transparent basis the scientific, technical and socio-economic information relevant to understanding the scientific basis of risk of human-induced climate change, its potential impacts and options for adaptation and mitigation. Review by experts and governments is an essential part of the IPCC process. The Panel does not conduct new research, monitor climate-related data or recommend policies. It is open to all member countries of WMO and UNEP.

In the United Nations Framework Convention on Climate Change (UNFCCC), explicit reference is made to the IPCC under Article 21:

> [T]he head of the interim secretariat referred to in paragraph 1 above will cooperate closely with the Intergovernmental Panel on Climate Change to ensure that the Panel can respond to the need for objective scientific and technical advice.

The IPCC was subsequently and repeatedly included in the Kyoto Protocol to the Convention. It was envisaged to take into account the methodological work of the IPCC within the framework of the Kyoto Protocol in formulating guidelines on verification of emission reductions.[4]

The IPCC consists of three working groups: The IPCC Working Group I (WG I) assesses the physical, scientific aspects of the climate system and climate change. The main topics assessed by WG I include: changes in greenhouse gases and aerosols in the atmosphere; observed changes in air, land and ocean temperatures, rainfall, glaciers and ice sheets, and oceans and sea level; historical and paleo-climatic perspectives on climate change; biogeochemistry, carbon cycles, gases and aerosols; satellite and

---

2 This chapter draws from the findings of the IPCC's Fifth Assessment report as well as from previous publications by the author, paricularly on Ruppel (2013), Ruppel & Ruppel-Schlichting (2016).

3 IPCC (2001).

4 See Article 3(4) of the Kyoto Protocol.

other data; climate models; and climate projections, causes and attribution of climate change.[5]

The IPCC Working Group II (WG II) assesses the vulnerability of socioeconomic and natural systems to climate change; identifies negative and positive consequences of climate change; and explores options for adapting to climate change. It also considers the relationship between vulnerability, adaptation and sustainable development. Information is evaluated by sector (water resources; ecosystems; food and forests; coastal systems; industry; human health) and region (Africa; Asia; Australia and New Zealand; Europe; Latin America; North America; Polar Regions; Small Islands).[6] In its reports, WG II elaborates on the scientific, technical, environmental, economic and social aspects of the vulnerability (sensitivity and adaptability) to climate change of, and the negative and positive consequences for, ecological systems, socioeconomic sectors and human health, with an emphasis on regional, sectoral and cross-sectoral issues.

The IPCC Working Group III (WG III) assesses options for mitigating climate change through limiting or preventing greenhouse gas emissions and enhancing activities that remove greenhouse gas from the atmosphere. The main economic sectors are taken into account, both from a short-term and a long-term perspective. The sectors include energy, transport, building, industry, agriculture, forestry, and waste management. WG III analyses the costs and benefits of the different approaches to mitigation, considering also the available instruments and policy measures. The approach is increasingly solution-oriented.[7]

The above-mentioned three working groups were intended to: [8]

> draw on slightly different scientific constituencies, since impact and responses would require factoring in research outside the physical sciences and would touch on political issues. Working Group I would be dominated by climate scientists, while Working Groups II and III would have a wider participation, including, as time went on, by economists and other social scientists.

Often, as in the IPCC, decisions among the dozens or hundreds of elite leaders are made by a negotiated consensus in a spirit of equality, of mutual accommodation, and of commitment to the community process – all

5 See IPCC on Working groups / Task Force at http://www.ipcc.ch/working_groups/working_groups.shtml, accessed 17 February 2013.

6 See http://www.ipcc.ch/working_groups/working_groups.shtml, accessed 17 February 2013.

7 See http://www.ipcc-wg3.de/, accessed 17 February 2013.

8 Mathiason & Bhandari (2010).

of which are seldom celebrated, but essential, components of the republican political culture.[9] It has been said that it is –[10]

> an important historical fact that such international regimes have been created chiefly by governments that felt comfortable with such mechanisms at home, that is, democratic governments. Nations like Nazi Germany, Communist China, and the former SU did little to create international organizations (aside from front groups under their own thumb), and often participated in them awkwardly. Happily, in the second half of the twentieth century, nations under democratic governance became globally predominant.

This encouraged the proliferation of international institutions that were democratic, or at any rate elite-based republican, exerting an ever-stronger influence in world affairs.[11] In 2007, the IPCC and Albert Arnold (Al) Gore Jr. were awarded the Nobel Peace Prize "for their efforts to build up and disseminate greater knowledge about man-made climate change and to lay the foundations for the measures that are needed to counteract such change".[12]

The IPCC is a very valuable institution that tries to help in an unprecedented way to resolve sociopolitical conflicts by gathering scientific knowledge and presenting it in a comprehensible manner. "The evidence shows the scientific consensus arrived at by the IPCC is a solid one, given the composition of the panel, and an innovative means of connecting science with politics."[13]

The Fifth Assessment Report was finalised in 2014. As had been the case in the past, the outline of the AR5 was developed through a scoping process which involved climate change experts from all relevant disciplines and users of IPCC reports, in particular representatives from governments. Governments and organisations involved in the Fourth Assessment Report were asked to submit comments and observations in writing with the submissions analysed by the panel. AR5 was delivered in stages, starting with the report of Working Group I on the physical science aspect, based on 9,200 peer-reviewed studies. The summaries for policy makers were released on 27 September 2013 for the first report, on 31 March 2014 for the second report entitled "Impacts, Adaptation, and Vulnerability", and on 14 April 2014 for the third report entitled "Mitigation of Climate

9 Weart (1998: 61).
10 Weart (2012).
11 Weart (1998: 262-267).
12 See http://www.nobelprize.org/nobel_prizes/peace/laureates/2007/, accessed 4 March 2013.
13 Mathiason & Bhandari (2010).

Change". The Synthesis Report was released on 2 November 2014, in time to pave the way for negotiations on reducing carbon emissions at the UN Climate Change Conference in Paris during late 2015.

The AR5 provides an update of knowledge on the scientific, technical and socioeconomic aspects of climate change. More than 800 authors, selected from around 3,000 nominations, were involved in writing the report.[14] In its reports the IPCC gives valuable advice to national governments and international organisations.[15] By effectively and objectively assessing scientific knowledge and prevailing uncertainty, the IPCC provides the world with the best possible and much-needed evidence of climate change-related impacts. Scientific authority also depends on reliable indicators.[16] In this context the IPCC plays – no doubt – a decisive role in the policy reform and political decision-making process:[17]

> Because of its scientific and intergovernmental nature, the IPCC embodies a unique opportunity to provide rigorous and balanced scientific information to decision makers. By endorsing the IPCC reports, governments acknowledge the authority of their scientific content. The work of the organization is therefore policy-relevant and yet policy-neutral, never policy-prescriptive.

The IPCC thus bridges the two fields, by getting the facts right so that the policies may be effective. This is important because "if scientists cannot agree, political leaders and other stakeholders are unlikely to agree either".[18]

C. Climate change, energy security and the relevance of bioenergy

The general message of AR5 can be summarised as follows: There is no doubt that we live in a world which is altered by climate change, one of the greatest challenges of the 21st century. Climate change poses risks to human and natural systems and has the potential to impose additional pressures on the various aspects of human security.[19] The risks and impacts related to climate change can be reduced by helping society to decrease vulnerability and bring down the overall risk level (adaptation)

---

14 See https://www.ipcc.ch/activities/activities.shtml, accessed 10 November 2016.
15 InterAcademy Council (2013).
16 Davis et al. (2012).
17 See http://www.ipcc.ch/organization/organization.shtml#.URelrmhpvos, accessed 17 February 2013.
18 Mathiason & Bhandari (2010: 58).
19 Adger & Pulhin (2014: 760).

and by reducing the amount of climate change that occurs. Thus, energy technologies play an important role in the field of climate change mitigation. Greenhouse gas emissions resulting from the provision of energy services have contributed significantly to the increase in atmospheric greenhouse gas concentrations and most – about 60% in 2010[20] – global anthropogenic greenhouse gas is attributed to the consumption of fossil fuels. Options for lowering greenhouse gas emissions from the energy system while still satisfying the global demand for energy services include energy conservation and efficiency, fossil fuel switching, nuclear and carbon capture and storage (CCS), and low greenhouse gas energy supply technologies such as renewable energy.

Energy security is one of the most important topics of our times. Energy is an essential requirement for all fields of our daily life, and for the functioning of social and political systems, businesses, and communication. So far, no commonly accepted definition of the term 'energy security' exists. In very general terms, energy security can be understood as robustness against disruptions of energy supply. The International Energy Agency (IEA) has defined energy security as "the uninterrupted availability of energy sources at an affordable price".[21] According to the IEA, energy security has a long-term component which –

> mainly deals with timely investments to supply energy in line with economic developments and environmental needs. On the other hand, short-term energy security focuses on the ability of the energy system to react promptly to sudden changes in the supply-demand balance.[22]

Energy security plays an important role at the crossroads of national security, economic security and environmental security, and is thus equally high on the agenda of national and international politicians, scientists and economists. To achieve secure, clean and efficient energy is the target of many national governments[23] and regional groups,[24] as

20 Victor et al. (2014: 122).

21 See the International Energy Agency at http://www.iea.org/topics/energysecurity/, accessed 7 November 2014. Of course, definitions of energy security vary and the definition proposed by the IEA is very broad and leaves questions unanswered, such as the question about what can be considered 'affordable' under the definition above.

22 International Energy Agency at http://www.iea.org/topics/energysecurity/, accessed 7 November 2014.

23 Namibia's Vision 2030, for example, sets out the objectives to achieve high value-added products and services; to provide security of energy supply through an appropriate diversity of economically competitive and reliable sources; to ensure that households and communities have access to affordable and appropriate ener-

securing energy supply is considered to be one of the means to overcome poverty and to achieve the Sustainable Development Goals (SDGs).

There is no question about the importance of energy security, in general; nor about its relevance for economic growth and development, in particular. However, in light of the fact that energy-related carbon dioxide emissions make up most global greenhouse gas emissions, the world community is charged with the task of balancing the extension of energy supply, on one hand, and the consumption of energy, on the other, in order to reduce the extent of climate change. Thus, international climate change negotiations are fundamentally about energy use and the linkages between energy and economic development.

To secure an energy supply that meets the growing demand is one of the major global energy challenges, and bioenergy plays an important role in this regard. Bioenergy is energy derived from biomass and can be deployed as solid, liquid and gaseous fuels. Bioenergy is suitable for a wide range of uses, including electricity production, transport, heating and cooking. On the one hand, bioenergy has a significant mitigation potential; on the other hand, there are issues to consider, such as the sustainability of practices and the efficiency of bioenergy systems. Bioenergy systems can cause a variety of positive and negative environmental, social, and economic consequences which need to be balanced when bioenergy is implemented.[25] Biomass-derived energy sources include wood and charcoal from forests (traditional forms of bioenergy), industrial burning of biomass wastes, the production of ethanol and biodiesel, and plantations of bioenergy crops (more modern forms of bioenergy). While traditional biofuels remain dominant energy sources in many less-developed parts of the

---

gy supplies; and to establish an energy sector that is efficient and that makes contributions to Namibia's economic competitiveness. Vision 2030 defines, as one of its strategies, the promotion of renewable energy sources and the implementation of projects for production from these sources to meet industry demand. See GRN (2004: 87).

24 The European Union (EU) for example "has agreed on ambitious Energy and Climate targets for 2020 and beyond to reduce greenhouse gas emissions, increase the share of renewable energies and improve energy efficiency. Achieving these objectives advances Europe along the path to an energy system that will deliver a competitive and secure energy supply which is sustainable." See EC (2014a). At a recent meeting of the European Council at the EU Summit in Brussels on 23 and 24 October 2014, EU leaders agreed to reduce greenhouse gas emissions by at least 40% compared to the 1990 level, and to increase energy efficiency and renewables by at least 27%.

25 Chum et al. (2011).

world, such as Africa, they have been in general decline in more developed countries as users switch to fossil fuels or electricity.

Box 1:
Definitions as contained in the IPCC's Fifth Asessment Report[26]

Bioenergy: Energy derived from any form of biomass such as recently living organisms or their metabolic by-products.

Bioethanol: Ethanol produced from biomass (e.g. sugar cane or corn).

Biofuel: A fuel, generally in liquid form, produced from organic matter or combustible oils produced by living or recently living plants. Examples of biofuel include alcohol (bioethanol), black liquor from the paper-manufacturing process, and soybean oil.

- First-generation manufactured biofuel is derived from grains, oilseeds, animal fats, and waste vegetable oils with mature conversion technologies.
- Second-generation biofuel uses non-traditional biochemical and thermochemical conversion processes and feedstock mostly derived from the lignocellulosic fractions of, for example, agricultural and forestry residues, municipal solid waste, etc.
- Third-generation biofuel is derived from feedstocks, such as algae and energy crops, by advanced processes still under development. Second- and third-generation biofuels produced through new processes are also referred to as next-generation or advanced biofuels, or advanced biofuel technologies.

The fact that energy is closely linked to climate change is reflected not only in the IPCC's AR5, but also in the Special Report on Renewable Energy Sources and Climate Change Mitigation (SRREN), which had already been published by the IPCC in 2012.[27] The SRREN has impartially assessed the scientific literature on the potential role of energy in the mitigation of climate change for policy makers, the private sector, academic researchers and civil society. The report furthermore considers the social and environmental consequences associated with the deployment of different energy technologies.

26 Allwood et al. (2014).
27 IPCC (2012).

Energy security and climate change share the need for innovation and technology, smart policy making, high levels of government attention, effective diplomacy, and international cooperation. Synergy effects will occur from innovative actions that make for a more secure energy system, as these may also result in reducing the warming emissions that come from energy supplies.[28]

The nexus between climate change and energy security has been focused on in the IPCC's assessment and it has been found that –

> most climate policies intersect with other societal goals, either positively or negatively, creating the possibility of 'co-benefits' or 'adverse side-effects'. Since the publication of AR4 a substantial literature has emerged looking at how countries that engage in mitigation also address other goals, such as local environmental protection or energy security, as a 'co-benefit' and conversely. This multi-objective perspective is important because it helps to identify areas where political, administrative, stakeholder, and other support for policies that advance multiple goals will be robust. Moreover, in many societies the presence of multiple objectives may make it easier for governments to sustain the political support needed for mitigation. Measuring the net effect on social welfare requires examining the interaction between climate policies and pre-existing other policies.[29]

It has been observed that the driving forces for climate policy are not solely the concern about climate change. This can be seen from the various efforts of national governments in which the issue of climate change is addressed in the context of other national objectives such as the alleviation of poverty and the achievement of energy security. "For countries that want to reduce their dependence on imported fossil fuels, climate policy can bolster energy efficiency and the domestic renewable energy supply, while cutting GHG emissions."[30]

The current over-reliance on finite traditional sources of energy, which are patchily distributed across the world's regions, has resulted in an intense competition over these resources. However, when looking at the ratio of proved reserves to current production, it is estimated that, globally, oil and natural gas will be exhausted in about four and six decades, respectively.[31]

Assessing the theoretical potential of renewable energy, the IPCC in its SRREN concludes that "the theoretical potential is much greater than all of the energy that is used by all the economies on Earth".[32] If supported by

28 See also World Economic Forum (2012).
29 Edenhofer et al. (2014: 40).
30 Kolstad & Urama (2014: 237).
31 See IPCC (2012: 122).
32 (Ibid.: 181).

the right enabling public policies, close to 80% of the world's energy supply could be met by renewables by mid-century.[33] While fossil energies are the engine of today's global economy,[34] renewable energies are one of the guarantors of a future-proof energy supply and they do play an important role in the fight against climate change. Wind power, solar energy, hydropower plants, geothermal energy, and energy from biomass are becoming important economic trends and key mechanisms to mitigate climate change. Technologies which supply energy with a limited production of greenhouse gases contribute to reducing the dependency on fossil fuels such as coal, oil and gas, and are constantly being developed further with a view to achieving a globally sustainable energy supply.

D. The international institutional and legal framework pertinent to energy security and climate change

Energy security is crucial for keeping economies competitive, for enhancing sustainable development, and for reducing poverty. It is thus, on the one hand, surprising that no global energy security system exists. On the other hand, it has to be taken into consideration that global energy governance must take into account a variety of topics, including climate change, development, environmental protection, trade, investment, and human security, which explains the fact that the global energy governance regime is fragmented, with its many components being managed in a disjointed manner, bringing about overlaps as well as normative gaps. Numerous instances of interstate cooperation interrelate and create a normative patchwork with implications for the global energy economy and security. This can be attributed to the pursuit of national interests, the diversity of energy sources, and the plurality of relevant institutions and agreements.[35]

Many international organisations are operating in the field of energy matters and include, among many others, the International Energy Agency (IEA); the Organization of the Petroleum Exporting Countries (OPEC); the Gas Exporting Countries Forum (GECF); the International Energy Fo-

33 IPCC (2011).

34 Within the 2012 estimates on energy share of global final energy consumption, fossil fuels amounted to 78.4%, nuclear energies 2.6% and renewables to 19% of the global share, REN21 (2014: 21).

35 For these and further aspects on the fragmentation of the global energy economy, see Leal-Arcas & Filis (2013).

rum (IEF); the World Trade Organization (WTO); and the International Renewable Energy Agency (IRENA).

A large number of international processes promote the acceleration of the deployment of renewable energies as an important part of designing a sustainable energy framework for the future. Among them are the Millennium Development Goals and the United Nations secretary general's Sustainable Energy for All Initiative, which is focused on three objectives, namely to ensure universal access to modern energy services (over one billion people worldwide lack access to electricity), to double energy efficiency (global energy-related carbon dioxide emissions could rise 20% by 2035; equipment maintenance, thermostat settings, and upgrades can reduce emissions by up to 50%), and to double the renewable energy share in the overall global energy mix by 2030 (global energy demand will grow to 33% from 2010 to 2035).[36]

Relevant instruments of a more legal nature include the UN Charter,[37] which sets out some foundation for international agreements relevant to energy-related issues by providing, for example, for the preservation of sovereignty over domestic matters, including the management of natural resources and especially of energy-related resources. Besides this, there are many other relevant international agreements, such as the Energy Charter Treaty and the legal regime of the WTO.[38]

Interrelating issues between climate change and energy security are most pertinently addressed within the regimes under the United Nations Convention on Climate Change (UNFCCC) and its Kyoto Protocol, which also address the environmental impacts of energy. The UNFCCC, as adopted in 1992, was designed to protect the climate system for present and future generations. It recognises that in order for developing countries to progress towards sustainable social and economic development –

> their energy consumption will need to grow taking into account the possibilities for achieving greater energy efficiency and for controlling greenhouse gas emissions in general, including through the application of new technologies on terms which make such an application economically and socially beneficial.[39]

Within the framework of the Kyoto Protocol, the enhancement of energy efficiency has been stipulated as a means for countries to achieve the

36 For more details on this initiative, see http://www.se4all.org, accessed 11 February 2015.

37 Available at http://www.encharter.org/fileadmin/user_upload/document/IEC/IEC_text_brochure_ENG.pdf, last accessed 30 January 2015.

38 On renewable energies, subsidies and the WTO, see Bougette & Charlier (2014).

39 See Preamble of the UNFCCC.

quantified emission limitation and reduction commitments. The 2015 Paris Agreement has confirmed such commitments. After 25 years of UN climate diplomacy, the world's governments have for the first time negotiated a treaty that envisages climate action by all countries. Ahead of the Paris negotiations, governments of 186 nations presented public plans (Intended Nationally Determined Contributions – INDCs) detailing how they would cut carbon emissions through to 2025 or 2030.[40] In this way countries set themselves targets for changing the direction of their economies toward climate-friendly development. The Paris Agreement thus enshrines in international law the target of limiting global warming to well below 2°C compared to pre-industrial levels and endeavours to limit it to 1.5°C. Alternative energy solutions will be key in achieving this ambitious goal.

Global energy transition will continue to be high on the international agenda in future, not only in terms of a new UN climate agreement, but also regarding a post-2015 agenda with universally applicable (applicable to all countries, not just developing nations and emerging economies), Sustainable Development Goals (SDGs), which were developed by the UN Open Working Group and adopted in 2015.[41] The goal pertinent to energy has been defined as follows:

> Goal 7. Ensure access to affordable, reliable, sustainable, and modern energy for all
>
> 7.1 by 2030 ensure universal access to affordable, reliable, and modern energy services
>
> 7.2 increase substantially the share of renewable energy in the global energy mix by 2030
>
> 7.3 double the global rate of improvement in energy efficiency by 2030
>
> 7.a by 2030 enhance international cooperation to facilitate access to clean energy research and technologies, including renewable energy, energy efficiency, and advanced and cleaner fossil fuel technologies, and promote investment in energy infrastructure and clean energy technologies
>
> 7.b by 2030 expand infrastructure and upgrade technology for supplying modern and sustainable energy services for all in developing countries, particularly LDCs and SIDS.[42]

40 For further information see the UNFCCC's Portal on INDCs at http://unfccc.int/focus/indc_portal/items/8766.php, accessed 10 December 2016.

41 Ruchser (2015); see also the Sustainable Development Knowledge Platform at https://sustainabledevelopment.un.org/sdgs, accessed 10 December 2016.

42 Available at https://sustainabledevelopment.un.org/focussdgs.html, accessed 30 January 2015.

E. The IPCC and focus topics related to biofuels production

While it is argued that the use of biofuels can be considered socially and economically beneficial and environmentally friendly as the CO_2 released when plants and trees are burned is balanced out by the CO_2 absorbed by the new ones planted to replace those harvested, the use of biofuels is also contested, with opponents of biofuels arguing that growing plants for energy puts pressure on land use, boosting deforestation and driving up food prices, and that the production of some biofuels has been linked to high levels of carbon emissions. The IPCC in its AR5 has addressed the pros and cons of biofuel production in many chapters, which shows the relevance of this topic related to the challenges posed by climate change.

Several benefits are associated with the increasing production of biofuels. In the first place, biofuels can potentially mitigate greenhouse gas emissions when used instead of fossil fuels such as gasoline, diesel, and more carbon-intensive fuels from tar sands and heavy oil. Furthermore, biofuels may boost investment, employment, and economic growth and offer supplemental livelihood opportunities. There are thus many convincing reasons for a country to encourage biofuel production. Biofuel production can also serve as a means to exert downward pressure on oil prices, to stimulate rural development and to reduce oil imports. Furthermore, second-generation biofuels, such as those based on non-food crops (e.g. grasses, algae and timber) and agricultural residues, are expected to produce reduced emissions of greenhouse gas and other air pollutants compared to most first-generation biofuels.[43]

Despite the aforementioned benefits of biofuel production, the effectiveness of biofuels as a mitigation strategy is contested and in several parts of the world evidence exists of the disruptive effect of biofuel production on food security, land tenure, labour rights and biodiversity. Owing to the potential negative consequences of its use as a mitigation strategy, bioenergy development leads to several emergent risks as shown in Table 1 below.

43 See Oppenheimer et al. (2014: 1055).

Table 1: Emergent risks related to biofuel production as a mitigation strategy[44]

Issue	Issue description	Nature of emergent risk	Reference
Direct and/or indirect land use change	Potential for enhancement of greenhouse gas emissions	Mitigation benefit of biofuels reduced or negated	Melillo et al. (2009a,b); Wise et al. (2009); Khanna et al. (2011)
Policies targeting only fossil carbon	Biofuel cropping competes with agricultural systems and ecosystems for land and water.	Mitigation benefit of policies reduced; harmful interactions with other key systems	Searchinger et al. (2008); Mellilo et al. (2009a,b); Wise et al. (2009); Fargione et al. (2008)
Food/fuel competition for land	Competition for land driving up food prices	Emergent risk of food insecurity due to mitigation-driven land use change	Searchinger et al. (2008); Pimentel et al. (2009); Hertel et al. (2010)
Biofuel production affects water resources.	Competition for water affects biodiversity and food cropping.	Emergent risk of biodiversity loss and food insecurity due to mitigation-driven water stress	Fargione (2010); Fingerman et al. (2010); Poudel et al. (2012); Yang et al. (2012)
Biofuel production affects biodiversity.	Competition for land reduces natural forest and biodiversity.	Emerging risk of biodiversity loss due to mitigation-driven land use change	Fizherbert et al. (2008); Koh et al. (2009); Lapola et al. (2010); Fletcher et al. (2011)
Land conversion causes air pollution.	Potential for increased production of tropospheric ozone from palm/sugarcane-induced land use change	Emergent risk of greenhouse gas-mitigation-driven plant and human health damage caused by tropospheric ozone	Cançado et al. (2006); Hewitt et al. (2009)
Fertilizer application	Potential for increased emissions of N_2O	Offsets some benefits of other mitigation measures	Donner and Kucharik (2008); Searchinger et al. (2008); Fargione (2010)
Invasive properties of biofuel crops	Potential to become an invasive species	Unintended consequences that damage agriculture and/or biodiversity	Raghu et al. (2006); DiTomaso et al. (2007); Barney and Ditomaso (2008)

44 Oppenheimer et al. (2014: 1056).

Areas vulnerable to bioenergy development include food security, land use, water security and biodiversity. The increasing cultivation of biofuels has various consequences. Changes in landscapes and increasing conflicts for scarce land resources are associated with large acquisitions of land for industrial agriculture, which may have negative consequences on local and marginalised communities. [45] Besides food price increases, biofuel production, often linked to transnational large-scale land acquisitions, raises another livelihood and poverty concern, namely the dispossession of land. In part as a result of climate change mitigation policies to promote biofuels, large-scale land acquisition in Africa has displaced small landholders. The greatest risk for negative outcomes for smallholders exists in Africa.[46]

Concerns exist that increased competition for scarce land undermines women's access to land and their ability to benefit economically from bio-fuel investment.[47]

An important aspect of food security is the prices of internationally traded food commodities. These prices reflect the overall balance of supply and demand, and the accessibility of food for consumers. Overall, food prices gradually declined for most of the 20th century. However, there have also been several periods of increases in international food prices. One major factor in recent price changes has been increased crop demand, which is ascribed to increased use of crops in biofuel production.[48]

The promotion of biofuel crops has been highly debated during the past years, as they have potential socioeconomic impacts for rural economies, they promote changes in land ownership and they affect food security. The IPCC's Special Report on Renewable Energy has emphasised that uncertainties exist around the role of biofuels in increases of food price and risks of deteriorating food security with future deployment of bioenergy.[49] However, increasing demand for biofuels shifts land from food to fuel production and has the potential to increase food prices, disproportionally affecting the poor. Poor urban residents are particularly vulnerable to food price increases as they use a large share of their income to purchase food. However, at the same time, higher food prices may benefit some self-employed agricultural groups.

---

45 Burkett et al. (2014: 180).
46 Daley & Englert (2010); Borras et al. (2011).
47 Daley & Englert (2010).
48 Porter et al. (2014: 494).
49 IPCC (2012: 273).

Concerns about the impact of biofuel production on food security not only relate to increases in food prices, but also to land concentration (and landgrabs), and to competition for water. Although there is high consensus that biofuel production does play a role in food prices, the extent of this influence remains disputed.

Given Africa's extensive arable lands, labour availability, and favourable climate for biofuel crop production, the continent has a significant potential for first-generation biofuel production derived from bioethanol from starch sources, and biodiesel production from oilseeds. On the positive side, it can be stated that biofuel production has positive energy security and economic growth implications. However, the prospect of wide-scale biofuel production in Africa also raises concerns regarding environmental and social sustainability. The following areas are considered to be most debated in terms of biofuel production:

- Competition for land and water;
- adverse impacts of biofuels on biodiversity and the environment;
- contractual and regulatory obligations that expose farmers to legal risks;
- changes in land tenure security; and
- reduced livelihood opportunities for women, pastoralists, and migrant farmers who depend on access to the land resource base.

Besides potential negative impacts on food security, biofuel schemes may also harm poor and marginalised people through declining biodiversity.

Water, energy, and food are interlinked through many interactive pathways affected by a climate change.[50] Like most energy production methods, the production of biofuels requires significant amounts of water. Biofuels produced from intensive agriculture will have an impact on rural areas and agriculture and aggravate stresses on water supplies, water quality and land use.[51] It has been reported that water for biofuels will drive global consumptive irrigation water use from 0.5% of global renewable water resources in 2005 to 5.5% in 2030, resulting in increased pressure on freshwater resources, with potential negative impacts on freshwater ecosystems.[52] Water for energy currently ranges from a few per cent in most developing countries to more than 50% of freshwater withdrawals in some developed countries. Future water requirements will depend on various

50 Field et al. (2014: 92).
51 Delucchi (2010).
52 Gerbens-Leenes et al. (2012).

factors including electricity demand growth, the portfolio of generation technologies and water management options employed.[53]

F. Concluding remarks

The IPCC informs not only the UNFCCC process with policy-relevant information, but also the governments of the world. Recent literature shows increasing attention to questions of ethics, justice, and responsibilities relating to energy security and climate change. Responses to mitigate climate change may thus have consequences and outcomes that favour certain populations or regions. Leaving aside the contested effectiveness of biofuels as a mitigation strategy, there is evidence of their disruptive effect on food security, land tenure, labour rights and biodiversity in several parts of the world. Climate policies encouraging cultivation of biofuels can encapsulate secondary impacts, both positive and negative. Their potential socioeconomic impacts are not to be underestimated, especially when it comes to water and food security, which has a particular impact on the poor. Competition over land has a long and bloody history in Africa too. Simultaneously, rising food and biofuel demand is a challenge that needs to be addressed when developing strategies to mitigate climate change.

That said, there are equally compelling reasons for a country to prohibit or encourage biofuels production depending on the strategy, policy and – of course – the country itself. More transdisciplinary research is therefore needed to better understand the socioeconomic and environmental tradeoffs associated with biofuel production in Africa. Without more effective policy frameworks to regulate the expansion of biofuels – and especially without a strong regulatory framework for land, investment management, and rural development – there is a risk that biofuels industrialisation could further exacerbate poverty and foster human insecurity in Africa.

Bibliography

Adger, WN & JM Pulhin, 2014, "Human security", in: *Climate change 2014: Impacts, aptation, and vulnerability, Part A: Global and sectoral aspects*, Contribution of Working Group II to the Fifth Assessment Report of the Intergovernmental Panel on Climate

53 Field et al. (2014: 92).

Change [Field, CB, VR Barros, DJ Dokken, KJ Mach, MD Mastrandrea, TE Bilir, M Chatterjee, KL Ebi, YO Estrada, RC Genova, B Girma, ES Kissel, AN Levy, S MacCracken, PR Mastrandrea & LL White (eds)], Cambridge, Cambridge University Press, at http://www.ipcc.ch/pdf/assessment-report/ar5/wg2/WGIIAR5-Chap12_FINAL.pdf, accessed 11 February 2015, 755.

Allwood JM, V Bosetti, NK Dubash, L Gómez-Echeverri & C von Stechow, "2014: Glossary", in *Climate change 2014: Mitigation of climate change,* Contribution of Working Group III to the Fifth Assessment Report of the Intergovernmental Panel on Climate Change [Edenhofer, O, R Pichs-Madruga, Y Sokona, E Farahani, S Kadner, K Seyboth, A Adler, I Baum, S Brunner, P Eickemeier, B Kriemann, J Savolainen, S Schlömer, C von Stechow, T Zwickel & JC Minx (eds)], Cambridge, Cambridge University Press, 1249.

Barney, JN & JM Ditomaso, 2008, Nonnative species and bioenergy: Are we cultivating the next invader?, *BioScience* 58(1), 64.

Borras Jr, SM, R Hall, I Scoones, B White & W Wolford, 2011, Towards a better understanding of global land grabbing: an editorial introduction, *The Journal of Peasant Studies* 38(2), 209.

Bougette P & C Charlier, 2014, Renewable energy, subsidies, and the WTO: Where has the 'green' gone? *GREDEG Working Paper* No. 2014-20.

Burkett, VR, AG Suarez, M Bindi, C Conde, R Mukerji, MJ Prather, AL St. Clair & GW Yohe, 2014, "Point of departure", in: *Climate change 2014: Impacts, adaptation, and vulnerability*, Part A: Global and sectoral aspects, Contribution of Working Group II to the Fifth Assessment Report of the Intergovernmental Panel on Climate Change [Field, CB, VR Barros, DJ Dokken, KJ Mach, MD Mastrandrea, TE Bilir, M Chatterjee, KL Ebi, YO Estrada, RC Genova, B Girma, ES Kissel, AN Levy, S MacCracken, PR Mastrandrea & LL White (eds)], Cambridge, Cambridge University Press, 169.

Cançado, JED, PHN Saldiva, LAA Pereira, LBLS Lara, P Artaxo, LA Martinello, MA Arbex, A Zanobetti & ALF Braga, 2006, The impact of sugar caneburning emissions on the respiratory system of children and the elderly, *Environmental Health Perspectives* 114(5), 725.

Chum H, A Faaij, J Moreira, G Berndes, P Dhamija, H Dong, B Gabrielle, AG Eng, W Lucht, M Mapako, OM Cerutti, T McIntyre, T Minowa & K Pingoud, 2011, "Bioenergy", in *IPCC special report on renewable energy sources and climate change mitigation* [O Edenhofer, R Pichs-Madruga, Y Sokona, K Seyboth, P Matschoss, S Kadner, T Zwickel, P Eickemeier, G Hansen, S Schlömer, C von Stechow (eds)], Cambridge, Cambridge University Press, 209.

Daley, E & B Englert, 2010, Securing land rights for women, *Journal of Eastern African Studies* 4(1), 91.

Davis, KE, B Kingsbury & SE Merry, 2012, Indicators as a technology of global governance, *Law & Society Review* 46(1), 71.

Delucchi, MA, 2010, Impacts of biofuels on climate change, water use, and land use, *Annals of the New York Academy of Sciences* 1195, 28.

DiTomaso, JM, JN Barney & AM Fox, 2007, Biofuel feedstocks: The risk of future invasions, All U.S. Government Documents (Utah Regional Depository), Paper 79, CAST Commentary QTA 2007-1, Council for Agricultural Science and Technology (CAST), at digitalcommons.usu.edu/govdocs/79, accessed 12 December 2016.

Donner, SD & CJ Kucharik, 2008, Corn-based ethanol production compromises goal of reducing nitrogen export by the Mississippi River, *Proceedings of the National Academy of Sciences of the United States of America* 105(11), 4513.

EC / European Commission, 2014, Draft Horizon 2020 work programme 2014-2015 in the area of secure, clean and efficient energy, at http://ec.europa.eu/research/horizon2020/pdf/work-programmes/secure_clean_and_efficient_energy_draft_work_programme.pdf, accessed 4 November 2014.

Edenhofer O, R Pichs-Madruga, Y Sokona, S Kadner, JC Minx, S Brunner, S Agrawala, G Baiocchi, IA Bashmakov, G Blanco, J Broome, T Bruckner, M Bustamante, L Clarke, M Conte Grand, F Creutzig, X Cruz-Núñez, S Dhakal, NK Dubash, P Eickemeier, E Farahani, M Fischedick, M Fleurbaey, R Gerlagh, L Gómez-Echeverri, S Gupta, J Harnisch, K Jiang, F Jotzo, S Kartha, S Klasen, C Kolstad, V Krey, H Kunreuther, O Lucon, O Masera, Y Mulugetta, RB Norgaard, A Patt, NH Ravindranath, K Riahi, J Roy, A Sagar, R Schaeffer, S Schlömer, KC Seto, K Seyboth, R Sims, P Smith, E Somanathan, R Stavins, C von Stechow, T Sterner, T Sugiyama, S Suh, D Ürge-Vorsatz, K Urama, A Venables, DG Victor, E Weber, D Zhou, J Zou, & T Zwickel, 2014, “Technical Summary”, in: *Climate Change 2014: Mitigation of Climate Change* – Contribution of Working Group III to the Fifth Assessment Report of the Intergovernmental Panel on Climate Change [Edenhofer, O, R Pichs-Madruga, Y Sokona, E Farahani, S Kadner, K Seyboth, A Adler, I Baum, S Brunner, P Eickemeier, B Kriemann, J Savolainen, S Schlömer, C von Stechow, T Zwickel & JC Minx (eds)], Cambridge, Cambridge University Press, at http://www.ipcc.ch/pdf/assessment-report/ar5/wg3/ipcc_wg3_ar5_technical-summary.pdf, accessed 11 February 2015.

Fargione, J, J Hill, D Tilman, S Polasky & P Hawthorne, 2008, Land clearing and the biofuel carbon debt, *Science* 319(5867), 1235.

Field, CB, VR Barros, KJ Mach, MD Mastrandrea, M van Aalst, WN Adger, DJ Arent, J Barnett, R Betts, TE Bilir, J Birkmann, J Carmin, DD Chadee, AJ Challinor, M Chatterjee, W Cramer, DJ Davidson, YO Estrada, JP Gattuso, Y Hijioka, O Hoegh-Guldberg, HQ Huang, GE Insarov, RN Jones, RS Kovats, P Romero-Lankao, JN Larsen, IJ Losada, JA Marengo, RF McLean, LO Mearns, R Mechler, JF Morton, I Niang, T Oki, JM Olwoch, M Opondo, ES Poloczanska, HO Pörtner, MH Redsteer, A Reisinger, A Revi, DN Schmidt, MR Shaw, W Solecki, DA Stone, JMR Stone, KM Strzepek, AG Suarez, P Tschakert, R Valentini, S Vicuña, A Villamizar, KE Vincent, R Warren, LL White, TJ Wilbanks, PP Wong & GW Yohe, “2014: Technical summary” in: Climate change 2014: Impacts, adaptation, and vulnerability. Part A: Global and Sectoral Aspects. Contribution of Working Group II to the Fifth Assessment Report of the Intergovernmental Panel on Climate Change [Field, CB, VR Barros, DJ Dokken, KJ Mach, MD Mastrandrea, TE Bilir, M Chatterjee, KL Ebi, YO Estrada, RC Genova, B Girma, ES Kissel, AN Levy, S MacCracken, PR Mastrandrea & LL White (eds)], Cambridge, Cambridge University Press, 35.

Fingerman, KR, MS Torn, MH O’Hare & DM Kammen, 2010, Accounting for the water impacts of ethanol production, *Environmental Research Letters* 5(1), 014020.

Fitzherbert, EB, MJ Struebig, A Morel, F Danielsen, CA Brühl, PF Donald & B Phalan, 2008, How will oil palm expansion affect biodiversity?, *Trends in Ecology and Evolution* 23(10), 538.

Fletcher Jr, RJ, BA Robertson, J Evans, PJ Doran, JRR Alavalapati & DW Schemske, 2011, Biodiversity conservation in the era of biofuels: risks and opportunities, *Frontiers in Ecology and the Environment* 9, 161.

Gerbens-Leenes, PW, AR van Lienden, AY Hoekstra & TH van der Meer, 2012, Biofuel scenarios in a water perspective: The global blue and green water footprint of road transport in 2030, *Global Environmental Change: Human and Policy Dimensions* 22(3), 764.

GRN / Government of the Republic of Namibia, 2004, Namibia vision 2030, Windhoek, Office of the President.

Hertel, TW, MB Burke & DB Lobell, 2010, The poverty implications of climateinduced crop yield changes by 2030, *Global Environmental Change* 20(4), 577.

Hewitt, CN, AR MacKenzie, P Di Carlo, CF Di Marco, JR Dorsey, M Evans, D Fowler, MW Gallagher, JR Hopkins, CE Jones, B Langford, JD Lee, AC Lewis, SF Lim, J McQuaid, P Misztal, SJ Moller, PS Monks, E Nemitz, DE Oram, SM Owen, GJ Phillips, TAM Pugh, JA Pyle, CE Reeves, J Ryder, J Siong, U Skiba & DJ Stewart, 2009, Nitrogen management is essential to prevent tropical oil palm plantations from causing ground-level ozone pollution, *Proceedings of the National Academy of Sciences of the United States of America* 106(44), 18447.

InterAcademy Council, 2013, IAC in brief, at http://www.interacademycouncil.net/23450/27799.aspx, accessed 17 February 2013.

IPCC / Intergovernmental Panel on Climate Change, 2001, Tenth anniversary brochure, at http://www.ipcc.ch/pdf/10th-anniversary/anniversary-brochure.pdf, accessed 17 February 2013.

IPCC / Intergovernmental Panel on Climate Change, 2011, Press release: Potential of renewable energy outlined in report by the Intergovernmental Panel on Climate Change, at http://srren.ipcc-wg3.de/press/content/potential-of-renewable-energy-outlined-report-by-the-intergovernmental-panel-on-climate-change, accessed 8 December 2014.

IPCC / Intergovernmental Panel on Climate Change, 2012, *Renewable energy sources and climate change mitigation*, Cambridge, Cambridge University Press, at http://www.ipcc.ch/report/srren/, accessed 10 November 2016.

Khanna, M, CL Crago & M Black, 2011, Can biofuels be a solution to climate change? The implications of land use change related emissions for policy, *Interface Focus* 1, 233.

Koh, LP, J Ghazoul, RA Butler, WF Laurance, NS Sodhi, J Mateo-Vega & CJA Bradshaw, 2010, Wash and spin cycle threats to tropical biodiversity, *Biotropica* 42(1), 67.

Kolstad C & K Urama, 2014, Social, economic and ethical concepts and methods, in: *Climate Change 2014: Mitigation of Climate Change*, Contribution of Working Group III to the Fifth Assessment Report of the Intergovernmental Panel on Climate Change [Edenhofer, O, R Pichs-Madruga, Y Sokona, E Farahani, S Kadner, K Seyboth, A Adler, I Baum, S Brunner, P Eickemeier, B Kriemann, J Savolainen, S Schlömer, C von Stechow, T Zwickel & JC Minx (eds)], Cambridge, Cambridge University Press, at http://www.ipcc.ch/pdf/assessment-report/ar5/wg3/ipcc_wg3_ar5_chapter3.pdf, accessed 11 February 2015.

Lapola, DM, R Schaldach, J Alcamo, A Boundeau, J Koch, C Koelking & JA Priess, 2010, Indirect land-use changes can overcome carbon savings from biofuels in Brazil, *Proceedings of the National Academy of the United States of America* 107(8), 3388.

Leal-Arcas, R & A Filis, 2013, The fragmented governance of the global energy economy: A legal-institutional analysis, *Journal of World Energy Law and Business* 6(4), 1, at http://papers.ssrn.com/sol3/papers.cfm?abstract_id=2296311, accessed 9 December 2014.

Mathiason, J & MP Bhandari, 2010, Getting the facts right: The intergovernmental panel on climate change and the new climate regime, *Journal of International Organization Studies* 1(1), 57.

Melillo, JM, AC Gurgel, DW Kicklighter, JM Reilly, TW Cronin, BS Felzer, S Paltsev, CA Schlosser, AP Sokolov & X Wang, 2009, Unintended environmental consequences of a global biofuels program, Report No. 168 *MIT Joint Program on the Science and Policy of Global Change*, Cambridge, at http://dspace.mit.edu/handle/1721.1/44626, accessed 15 December 2016.

Melillo, JM, JM Reilly, DW Kicklighter, AC Gurgel, TW Cronin, S Paltsev, BS Felzer, X Wang, AP Sokolov & CA Schlosser, 2009, Indirect emissions from biofuels: how important?, *Science* 326(5958), 1397.

Oppenheimer, M, M Campos, R Warren, J Birkmann, G Luber, B O'Neill & K Takahashi, 2014, "Emergent risks and key vulnerabilities", in: *Climate change 2014: Impacts, adaptation, and vulnerability*, Part A: Global and Sectoral Aspects, Contribution of Working Group II to the Fifth Assessment Report of the Intergovernmental Panel on Climate Change [Field, CB, VR Barros, DJ Dokken, KJ Mach, MD Mastrandrea, TE Bilir, M Chatterjee, KL Ebi, YO Estrada, RC Genova, B Girma, ES Kissel, AN Levy, S MacCracken, PR Mastrandrea & LL White (eds)], Cambridge, Cambridge University Press, 1039.

Pimentel, D, A Marklein, MA Toth, MN Karpoff, GS Paul, R McCormack, J Kyriazis & T Krueger, 2009, Food versus biofuels: environmental and economic costs, *Human Ecology* 37(1), 1.

Porter, JR, L Xie, AJ Challinor, K Cochrane, SM Howden, MM Iqbal, DB Lobell & MI Travasso, 2014, "Food security and food production systems", in: *Climate change 2014: Impacts, adaptation, and vulnerability*, Part A: Global and Sectoral Aspects. Contribution of Working Group II to the Fifth Assessment Report of the Intergovernmental Panel on Climate Change [Field, CB, VR Barros, DJ Dokken, KJ Mach, MD Mastrandrea, TE Bilir, M Chatterjee, KL Ebi, YO Estrada, RC Genova, B Girma, ES Kissel, AN Levy, S MacCracken, PR Mastrandrea & LL White (eds)], Cambridge, Cambridge University Press, 485.

Poudel, BN, KP Paudel, G Timilsina & D Zilberman, 2012, Providing numbers for a food versus fuel debate: An analysis of a future biofuel production scenario, *Applied Economic Perspectives and Policy* 24(4), 637.

Raghu, S, RC Anderson, CC Daehler, AS Davis, RN Wiedenmann, D Simberloff, & RN Mack, 2006, Adding biofuels to the invasive species fire?, *Science* 313(5794), 1742.

REN21 / Renewable Energy Policy Network for the 21st Century, 2014, Renewables 2014 global status report, at http://www.ren21.net/portals/0/documents/resources/gsr/2014/gsr2014_full%20report_low%20res.pdf, accessed 11 February 2015.

Ruchser, M, 2015, Where is the energy transition headed in 2015?, *The Current Column*, German Development Institute (DIE), 23 January 2015, at http://www.die-gdi.de/uploads/media/German_Development_Institute_Ruchser_23.01.2015.pdf, accessed 30 January 2015.

Ruppel, OC, 2013, "Intersections of law and cooperative global climate governance – challenges in the anthopocene", in: Ruppel, OC, C Roschmann & K Ruppel-Schlichting (eds), *Climate change: International law and global governance*, Volume I: Legal responses and global responsibility, Baden-Baden, Nomos, 29.

Ruppel, OC & K Ruppel-Schlichting, 2016, "Comparative legal aspects of the potential of renewable energies to promote energy security, sustainable development and climate change mitigation", in: Ruppel, OC & B Althusmann (eds), *Perspectives on energy security and renewable energies in sub-Sahara Africa – Practical opportunities and regulatory challenges*, Windhoek, MacMillan Education Namibia, 117.

Searchinger, T, R Heimlich, RA Houghton, F Dong, A Elobeid, J Fabiosa, S Tokgoz, D Hayes & T Yu, 2008, Use of US croplands for biofuels increases greenhouse gases through emissions from land-use change, *Science* 319(5867), 1238.

Victor DG, D Zhou, EHM Ahmed, PK Dadhich, JGJ Olivier, HH Rogner, K Sheikho & M Yamaguchi, 2014, "Introductory Chapter", in: *Climate Change 2014: Mitigation of Climate Change*, Contribution of Working Group III to the Fifth Assessment Report of the Intergovernmental Panel on Climate Change [Edenhofer, O, R Pichs-Madruga, Y Sokona, E Farahani, S Kadner, K Seyboth, A Adler, I Baum, S Brunner, P Eickemeier, B Kriemann, J Savolainen, S Schlömer, C von Stechow, T Zwickel & JC Minx (eds)], Cambridge, Cambridge University Press, at http://www.ipcc.ch/pdf/assessment-report/ar5/wg3/ipcc_wg3_ar5_chapter1.pdf, accessed 11 February 2015.

Weart, SR, 1998, *Never at war: Why democracies will not fight one another*, Yale, Yale University Press.

Weart, SR, 2012, The evolution of international cooperation in climate science, *Journal of International Organization Studies* 3(1), 41.

Wise, M, K Calvin, A Thomson, L Clarke, B Bond-Lamberty, R Sands, SJ Smith, A Janetos & J Edmonds, 2009, Implications of limiting CO_2 concentrations for land use and energy. 324 (5931) *Science*, 1183.

World Economic Forum, 2012, *White paper on energy security and global warming*, at http://www3.weforum.org/docs/WEF_GAC_WhitePaperEnereySecurityGlobalWarming_2012.pdf, accessed 5 December 2014.

Yang, Y, J Bae, J Kim & S Suh, 2012, Replacing gasoline with corn ethanol results in significant environmental problem-shifting, *Environmental Science and Technology* 46(7), 3671.

Socio-enviro-economic impacts of biofuels: Weighing the risks and benefits for southern Africa

Manisha Gulati, BA (Hons) Economics and Masters in Business Economics (University of Delhi, India), Energy Economist with the World Wide Fund for Nature (WWF), South Africa

Louise Scholtz, BA LLB MBA and MPhil in Sustainable Development (University of Stellenbosch), Policy and Futures Unit, World Wide Fund for Nature (WWF), South Africa

A. Introduction

Biofuels are increasingly seen as the panacea for a whole range of challenges that Africa faces. The African Union regards biofuels as an "integral part of the sustainability energy strategy for the continent" and the eradication of poverty on the continent. It has urged countries to develop their own biofuel strategies aligned to their natural resources, land tenure and socio-economic needs.[1] In line with this positive endorsement, southern African policy makers are at the various stages of creating a favourable environment to attract investments in biofuels and initiating commercial production of biofuels. Their motivations range from diversification of energy sources, reducing dependency on imported oil and shifting away from dependence on biomass sources such as charcoal and wood, to reducing greenhouse gas (GHG) emissions in the transport sector, and the stimulation of rural development through job creation.

Large scale new technologies, including the introduction of new crops as biofuels, raise concern about potential impacts on society – ecological, environmental and economic – that will not be captured if success is measured only in terms of money. In addition public policies have effects that "are inadvertent or unanticipated".[2] Wolde-Georgis & Glantz further alert that current biofuel practices and strategies in Africa undermine the

1 Wolde-Georgis & Glantz (2008: 5).
2 (Ibid.: 11).

"social and environmental equity promises of biofuels development",[3] pointing to a profound disconnect between the stated intent of policies and practice.

Jambe & Mkondiwa put the dilemma succinctly as follows:[4]

> Of particular eminence are the three dilemmas that policy makers especially in developing countries face when considering the development of the biofuels sector. These regard the questions as to whether biofuels production and use implies a choice between food and fuel; whether biofuels have positive or negative effects on climate change and the broader environment; or whether biofuels contribute to the socioeconomic development, wealth creation and distribution.

Potential positive and negative outcomes depend to a large extent on the development strategy chosen. Essentially there are three options: In the first instance large scale plantations that create a vertically integrated production chain, secondly an option that allows local farmers to enter into contracts with refineries to supply feedstock; and thirdly, the cultivation by small scale farmers for processing at local level. All three carry risks. Large-scale plantations are often created without the necessary land use and environmental impact assessments and bypass local farmers that might result in farmers being displaced and impoverished. The second option, by comparison might offer benefits to local farmers, but require government oversight and insurance mechanisms to buffer farmers against biofuel market failures, unscrupulous practices by investors or biofuel companies that go bankrupt. In the absence of these interventions, farmers run the risk of being confronted by food shortages, famine or even loss of their farms. The last option probably offers the most benefits, providing farmers with new farming and processing skills and not alienating them from their lands. However, there might be food security issues with farmers moving away from food cultivation to biofuel cultivation.[5]

Accordingly policies should address a multitude of concerns – protection of the rights of local farmers and landowners, ensure developmental benefits, address food security concerns, and mitigate against potential negative impacts on the environment. However, southern Africa has a patchy record when it comes to enforcing environmental legislation aimed at protecting critical resources such as land and water. An added area of concern is the complicity of local elites in land 'grabbing' by large produces to grow biofuels. Of utmost importance in this regard would be

---

3 Wolde-Georgis & Glantz (2008: 17).
4 Jambe & Mkondiwa (2013: 614).
5 Wolde-Georgis & Glantz (2008: 11).

strong government support for small biofuel producers to pre-empt commercial interests hijacking the process and steering production to large scale schemes that arguably provide better returns on investment.[6]

This paper argues that existing policies do not always display an understanding of the socio-economic and environmental impacts of biofuels. Biofuel production sits at the critical intersection between food, water and energy – also known as the food-water-energy nexus. The pursuit of biofuel production in Africa will entail trade-offs, balancing the need to address food security in a continent with the need to create jobs and livelihoods as well as access to energy. Furthermore, water is a critical resource in much of Africa, and should be allocated optimally.

Understanding these interconnected issues is important for a thorough assessment of the true opportunities and risks of biofuel production. This paper aims to contribute to a better understanding of these impacts, thereby contributing to policy options that governments may wish to pursue. The objective is to encourage policy-making that appreciates and grasps the complexities of biofuel production and consumption and balances them so as to avoid unintended consequences.

The next section (B), which is the heart of this paper, provides an overview of the dominant socio-environmental and economic problems and perspectives toward biofuel production. Section C discusses whether biofuels will actually contribute to sustainable energy access, reduced energy poverty and rural economic development in Africa and live up to the promise of delivering rural development benefits. Section D outlines the status of policy development for biofuels in southern Africa including the drivers of such policy-making, while Section E closes the paper with recommendations for policy-making.

B. Overview of the dominant socio-environmental and economic problems

Biofuel production will entail both positive and negative socio-environ-economic impacts, with trade-offs between the different impacts. The potential impacts include uncertainty in quantum of GHG emissions reduced, changes in land use, increased water use and the impacts of increased nutrient and pesticide applications. Similarly, conversion into biofuels and subsequent end use of the fuel will also have wider environmental impacts

6 Ewing & Msangi (2009: 520).

including water usage and contamination and air quality. It is also important to consider the indirect environmental impacts that might arise as a result of the interactions between different land uses, such as land used for biofuels, food and material production.

First, and in respect of GHG emissions and air quality, a variety of studies have shown that life cycle GHG savings vary significantly (see Figure 1) depending on type of land used, choice of feedstock, geo-climatic conditions, feedstock cultivation methods and conversion technologies.[7] While first generation biofuels reduce emissions by 20 to 60%, second generation biofuels typically reduce emissions by 70 to 90%.[8] When production takes place on converted natural land meaning by converting forests and wetlands into cropland, emissions may be higher than that from fossil fuels and the carbon debt created from conversion of natural land may not be 'paid back', even over a long period of time.[9] GHG emissions can also vary for the same feedstock based on site-specific conditions, processing methods, and fuel selection (see Figure 2).

Figure 1: GHG savings from biofuels compared to fossil fuels[10]

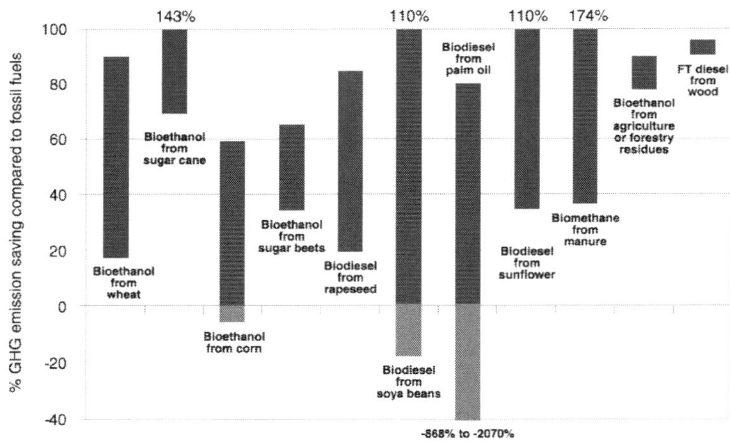

7 See for example FAO (2008); Wang et al. (2007); Reinhardt et al. (2008).
8 FAO (2008).
9 Fargione et al. (2008).
10 Source: UNEP (2009).

Figure 2: Well-to-wheels GHG emissions changes for corn ethanol under various processing scenarios, shown in comparison with cellulosic ethanol (comparisons on energy equivalent basis)[11]

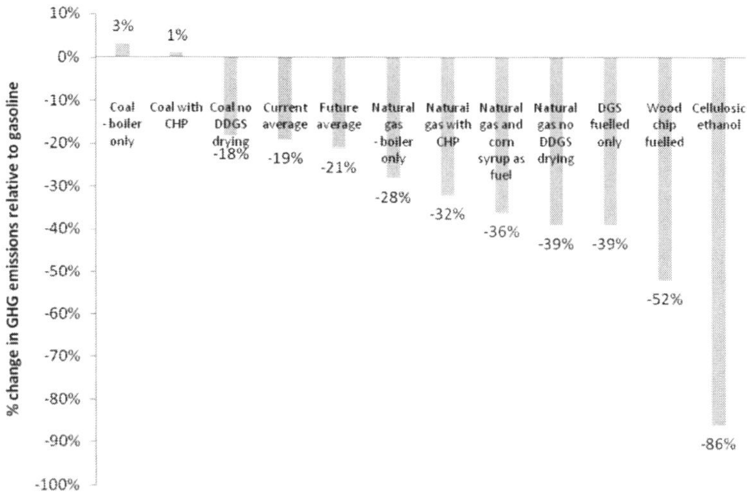

Second, many environmental impacts of biofuels are generally not accounted for. Biofuels tend to have impacts on ozone depletion, acidification, and eutrophication[12] (see Figure 3).[13] While biofuels have lower sulphuric emissions compared to fossil fuels, they generate higher nitrogen oxide (NOx) emissions in comparison to fossil fuels. Biodiesel, for example, can generate up to 70% higher NOx emissions depending on feedstock. Displacement of gasoline by some high ethanol blends could also lead to greater emission of local air pollutants and negative health impacts from deteriorating air quality.[14]

11 Source: Kunen & Chalmers (2010).

12 'Eutrophication' is generally defined as an increase in nutrients such as nitrogen and phosphorus that increase algal growth. Depending on the degree of eutrophication, severe environmental effects can develop, which degrade water quality.

13 UNEP (2009).

14 Jacobson (2007).

Figure 3: Environmental impacts of biodiesel from rapeseed compared to conventional diesel fuel[15]

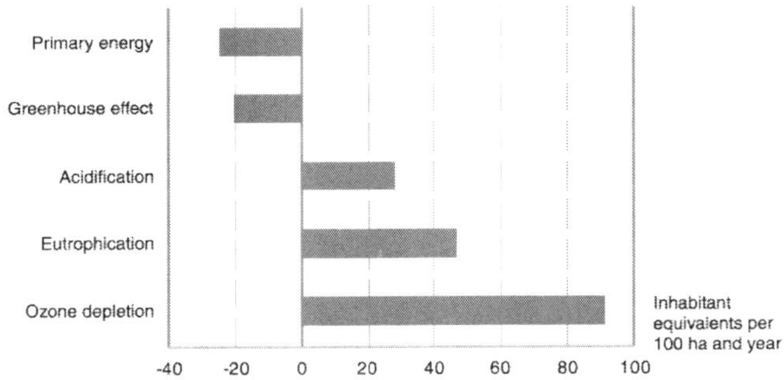

Third, the land requirement of different feedstock to produce the same quantum of biofuels can vary significantly (see Figure 4). The estimates of land requirements for biofuels also vary widely (see Table 1) and depend on assumptions for type of feedstock, geographical location, level of input and yield increase. Diversion of vast amounts of cropland to meet the land requirements of biofuels production can pose a threat to food production. In addition, the demand for land can prompt large-scale land conversion and result in loss of biodiversity and increase in forest fires besides negating the emissions savings from biofuel development.

This conversion may hold significant carbon implications. The suitability of many current biofuel crops to Africa's tropical climate creates an incentive to convert natural ecosystems into plantations or peatlands. Although computing total emissions is complex due to the difficulty of bounding the system in terms of computing the carbon balance of biofuel systems, compared to conventional fuel systems,[16] estimates suggest that converting forests, peatlands, grasslands or savannas to produce first generation biofuels is likely to create a carbon debt with 17 to 420 times more emissions than annual emissions reductions from biofuels displacing fossil fuels.[17] Finally, large scale expansion of biofuels could also result in loss of agro-biodiversity due to the intensification of agriculture through mono-cropping since most biofuel feedstock plantations rely on a single species.

15 Reinhardt et al. (2008).
16 von Maltitz & Brent (2008).
17 Fargione et al. (2008).

Figure 4: Land required by different biofuels feedstock for producing the same amount of energy[18]

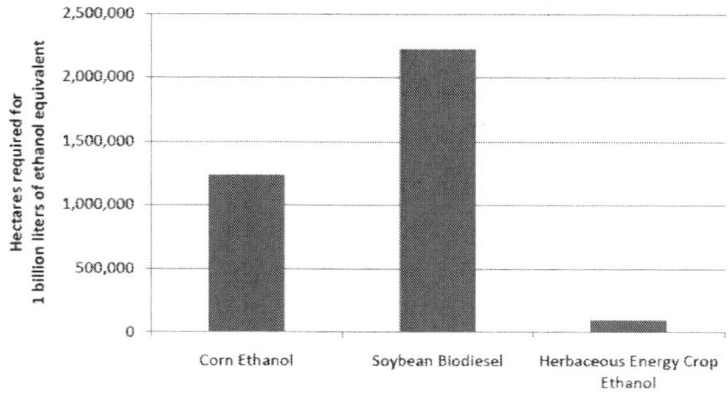

Table 1: Projections of land requirements for biofuels development[19]

	Target	Cropland requirement
IEA Reference Case (2006)	3% of transport energy in 2030	34.5 MHa
OECD / FAO (2007)	Projected increase until 2020	35 Mha
Renewable Fuels Agency 2008	All major countries and regions to achieve targets	56 to 166 Mha
Ravindranath et al. (2009)	10% supply of global transport fuel demand in 2030 with 1G biofuels	118-508 Mha

Such land conversion is already being experienced in Indonesia and some parts of Africa where the demand for palm oil is resulting in conversion of forested areas into palm oil plantations. Two-thirds of the current expansion of palm oil cultivation in Indonesia is based on the conversion of rainforests, with one third planted on previously cultivated or fallow land. If current trends continue, the total rainforest area of Indonesia would be reduced by 29% as compared to 2005, and would only cover about 49% of the original area from 1990. Moreover, huge forest fires in 1997 and 1998 destroyed 10 Mha of rainforest area in Borneo, Sumatra and New-Guinea,

18 Source: Kunen & Chalmers (2010).
19 Source: UNEP (2009).

boosting global atmospheric CO_2 concentrations in 1997 to almost twice the average values of years before and after 1997.

In Ethiopia, the Biofuels Development and Utilization Strategy is encouraging large scale production of biofuels without conducting proper land inventories. For instance, Flora Eco Power Ethiopia, operated by a German private company, purchased 200,000 ha to plant castor seeds for biodiesel production. By 2008, they had cultivated 15,000 ha in several Woredas in East and West Hareghe Zones and cleared 10,000 ha of virgin forestland. Environmental impact assessments are either not required before such clearance or are performed in an ad hoc manner.[20]

The situation in southern Africa mirrors the examples of elsewhere with pressures mounting to commercialise land that has resulted in land being transformed and overtaken by the widespread leasing or sale of public lands to foreign companies and governments. In some instances for biofuel production, this dates to the oil price spikes of the mid-2000s. In fact it is the rapid expansion of investments to secure land to produce biofuels was what initially focussed attention on the rush for southern Africa's farmland, partly driven by companies seeking to meet the demand for renewable fuel stocks in the North, and the European Union in particular (See Figure 5).[21]

Several countries, including Germany, Israel, India and the USA have already entered into investment agreements with African countries and acquired land, and in some instances started producing biofuels. This interest notwithstanding, the industry in Africa is still in its infancy, with all parts of the world expecting Latin America (excluding Brazil), producing more biofuels annually.[22]

20 Amigun (2011).
21 Hall (2011: 3).
22 Wolde-Georgis & Glantz (2008: 4).

Figure 5: Expected share of global biofuel consumption in 2030 for main consumer markets[23]

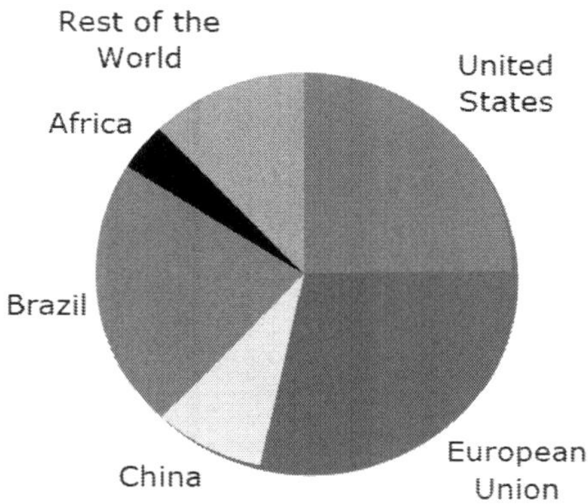

A possible solution to prevent large-scale land conversion and to realise beneficial effects for biodiversity is to encourage biofuel production on degraded or marginal lands (see Figure 6). But degraded lands by definition are ill suited for agriculture, typically lacking water and nutrients and requiring higher inputs. A trade-off here would be that biofuels production on abandoned land is likely to reduce less emissions than re-establishing natural forest on such land.

23 Source: IEA (2010).

Figure 6: Short-term impact of land conversion[24]

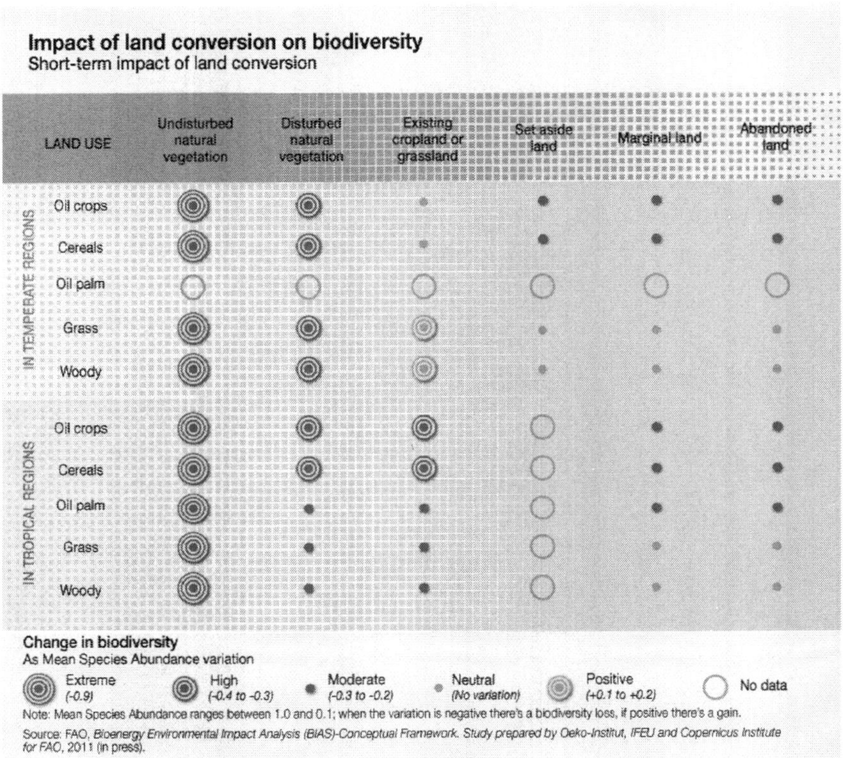

Fourth, both production and processing of some biofuels requires large volumes of water, which is problematic in semi-arid areas. The water requirement for the same biofuel feedstock may vary significantly depending on region (see Figure 7). In addition, the water requirement by biofuels can be higher than that from other sources of energy. One study[25] estimated that biofuels can require 70 to 400 times more water than fossil fuels, wind and solar energy. Another study estimated that roughly 45 billion cubic meters of irrigation water was used for biofuels production in 2007, some 6 times more water than used for drinking water globally.[26]

---

24 Source FAO (2011).
25 Gerbens-Leenes et al. (2009).
26 UNEP (2009: 56).

Figure 7: Water footprint analysis of corn ethanol in three regions in the United States based on irrigated water use[27]

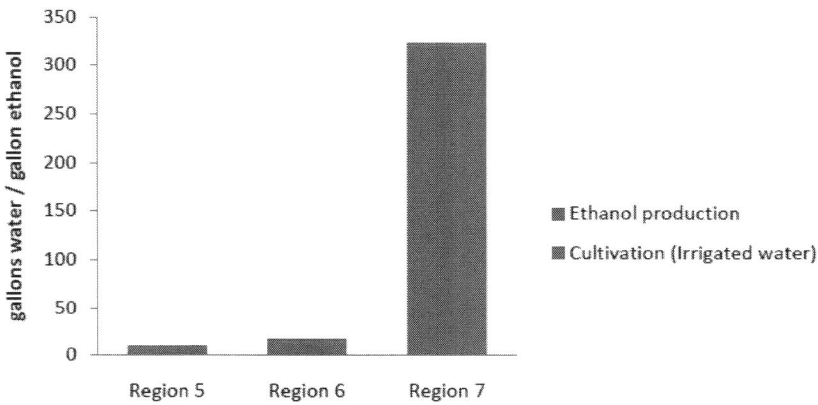

With many regions of southern Africa being water scarce and with little rainfall, feed-stock production for biofuels will require irrigation to be commercially viable. Given that water requirements for biofuels will be considered under water allocation for agriculture, it is likely that the high water requirements of biofuels could constrain water availability for food production. Biofuels also risk exacerbating water scarcity, which can pose a serious risk for local communities, especially in highly populated areas, leading to conflicts and impacting on local self-sufficiency. In order to minimise potential risk of conflict, biofuel investors often target less populated land, where the size of communities are smaller coupled with weak law enforcement and low media exposure make it easier to manage the conflict.[28]

Research at the International Water Management Institute (IWMI) in Sri Lanka has shown that, to produce one litre of biofuel between 1,000 and 4,000 litres of water is evaporated, depending on the feedstock and conversion techniques.[29] In fact, a study conducted by the International Water Management Institute on impact of biofuels in China suggests that the potential strain of biofuels on water resources may be enough to persuade policy makers to abandon biofuel production altogether.[30]

---

27 Source: Kunen & Chalmers (2010).
28 Sielhorst et al. (2008).
29 Nadia (2009).
30 De Fraiture et al. (2008: 67).

Finally, there is the issue of trade-offs with food production. The impact of biofuels on food security and prices is complex. However, serious concerns have been expressed regarding the possible role of biofuels in agro-inflation in the 2008 food crisis. Worldwide biofuels production growth over 2 years ending June 2008 is estimated to have accounted for 12% of the rise in IMF's food price index, with roughly 60%, 14% and 15% of this 12% increase being attributed to increased biofuels production in the US, Brazil and the EU respectively.[31] A study by IFPRI in 2009 indicated that using crops for biofuel use led to an increase of 25 to 30% in food prices.[32] IFPRI also projects price hikes for maize of 23 to 72%, wheat of 8 to 30%, oilseeds of 18 to 76% and sugar of 11.5 to 66%, with countries implementing 2020 biofuels plans.

Some studies concur that the diversion of the US corn crop to biofuels production was the strongest demand-induced force on food prices in 2008, given that the US accounted for about one-third of global maize production and two-thirds of global exports. The expansion of maize area in the US by 23% in 2007 entailed the contraction of soybean area by 16%, leading to lower soybean output and playing a part in the 75% rise in soybean prices from April 2007 to April 2008.

The increased demand for such food crops as resources for biofuels will lead to rise in prices for these crops. Besides availability and price of food, there are concerns around nutrition. One estimate suggests that if the current biofuel expansion continues with respect to the various targets set up by different nations, the calorie availability in developing countries is expected to grow more slowly and the number of malnourished children is expected to increase even more.[33] In addition, competition over land and water will also grow with the increased demand for biofuel production.

It is to avoid these consequences that the South African biofuels strategy has excluded the use of maize as a feedstock for ethanol production. South Africa further provides for bio-ethanol production from maize to be considered only once certainty on the ability of the currently underutilised land to produce has been ascertained and the necessary measures are in place to guard against food inflation. Although provisions such as this one are good from the food security perspective, it needs to be noted that there is nothing to prohibit farmers to move away from crops that are key to food security to energy crops such as sugar cane or sunflower. If farmers

31 Baier et al. (2009).
32 BirdLife International (2009).
33 Rosegrant (2008); Barbara (2007).

find a good price for their energy crop from biofuels producers or if the market prices for the non-food and energy crops go up, there would be natural inclination to shift towards energy crops. In that case, concerns of price hikes and supply of basic food crops would still prevail.

Table 2: Estimated impact of biofuels on food prices[34]

Study	Coverage	Key impacts of biofuels on food prices
Banse et al. (2008)	2001 to 2010	Price change under reference scenario, 5.75% blending, and 11.5% blending, respectively: Cereals: -4.5%, 1.75%, +2.5% Oilseeds: -1.5%, +2%, +8.5% Sugar: -4%, -1.5%, +5.75%
Baier et al. (2009)	24 months ending June 2008	Global biofuel production growth responsible for 17%, 14% and 100% of the rises in corn, soybean and sugar prices, respectively.
IMF (2008)	12 months ending March 2008	25-45% of the rise in corn prices attributable to ethanol production increase in US.
Lipsky (2008) & Johnson (2008)	2005 to 2007	Increased demand for world biofuels accounts for 70% of the increase in corn prices.
Rosegrant et al. (2008)	2000 to 2007	Increased biofuel demand is found to have accounted for 30% of the increase in weighted average grain prices, 39% of the increase in real maize prices, 21% of the increase in rice prices and 22% of the rise in wheat prices.

C. *Will biofuels live up to their promise in Africa?*

As mentioned earlier, many African policy makers are viewing biofuels as a route to access to affordable and modern energy sources, reduced energy poverty, rural economic development, employment, poverty reduction and even and improved foreign exchange reserves. However this vision of a rural bio-economy would have to be accompanied by broader investment in rural infrastructure, to have multiplier effects. Other barriers to the

34 Source: Timilsina & Shrestha (2010).

development of a rural biofuel economy relate to lack of technical skills, available start-up capital, a dearth of private sector capacity and support, supply chain development, and land tenure arrangements.[35]

Managed well, biofuels can pave the way for long term agrarian transformation, through economically and environmentally sustainable (integrated) cropping and sound policy and institutional interventions as part of the investment in biofuel value chains that form part of a larger agricultural system, with subsequent benefits to rural communities.[36] Theoretically, these postulations are reasonable. In reality, whether biofuels live up to this ambition is complex and difficult to generalise.

Arguably, domestically grown biofuels may become an attractive substitute for imported fossil fuels for many African countries and in so doing impact positively on energy security. A 2004 IEA study found that a 10 USD increase in oil prices meant 1.5% lower GDP for oil importing African countries. In 2000, sub-Saharan African countries spent 14% of their income on oil imports.[37] However, the blending targets in southern African countries are too low to alleviate energy security concerns around liquid fuels either on quantity or price front. For example, South Africa imports the majority of its oil. Its target of 2% penetration level of biofuels in national liquid fuels pool amounting to 400 million litres per annum. This is unlikely to dramatically increase the country's energy security as far as oil consumption is concerned.

As for reduced expenditures for oil import, which has been extremely high in the past due to high oil prices, the net gain for any country would depend on the savings in oil imports and the investments in infrastructure for biofuels. However, oil price volatility calls into question the economic viability of large agrofuel initiatives. Cheaper oil, coupled with increased costs of producing, refining and transporting biofuels have led to a waning in the initial rush for land suitable for biofuels.

However, political support for the pursuit of biofuels as an integral part of regional energy strategies, have not abated, with Zimbabwe a case in point (see Box 1).

35 Ewing & Msangi (2009).
36 Brent (2014).
37 Sielhorst et al. (2008).

Box 1: Zimbabwe Case study[38]

A jatropha processing plant was set up in 2007 in Mount Hampden, Harare with the capacity to produce 100 million litres of biodiesel annually – the first such commercial plant on the continent. The plant was jointly owned by the Reserve Bank of Zimbabwe (RBZ) and private investors. The fuel crisis crippling the economy at the time due to lack of foreign currency was the major driving force that led to the resuscitation of the programme. Estimations at the time predicted savings of close to US$ 80 million annually from diesel importation. The plant produced over 500,000 litres of bio-diesel in 2008 and almost 200,000 litres in 2009 using cotton seed supplied by Cargill and Cottco as feedstock. However, this was less than 5% its output capacity and in April 2010, the government stopped funding it. The main reasons put forward for the failure of the project was lack of coordination in running the programme and the inability to pay an attractive price to the farmers for jatropha which eventually led to lack of constant and reliable supply of the feedstock.

This, however, did not discourage the Zimbabwean government and in 2011 a large bioethanol plant project was resuscitated at Chisumbanje in the Chipinge District. The plant was set up on a joint venture partnership between the governmental Agricultural Rural Development Agency (ARDA) and a private investor, Green Fuels, once again to be the largest of its kind in sub-Saharan Africa. It was envisaged that when fully operational, the plant will produce 700,000 litres of fuel daily equalling 27% of the country's fuel needs, achieving foreign currency savings in the order of 100 million USD per year (depending on international fuel prices). In 2011 Green Fuels started producing anhydrous ethanol and began selling ethanol on the market. During the third quarter of the year, FMI Energy (BP) began to dispense blended petrol at their retail sites. The bioethanol production facility was expected to create thousands of jobs, lower local fuel prices and also improve the livelihood of rural folk living within the vicinity of the plant. Unfortunately, in February 2012, the plant shut down its operation due to numerous challenges, including a dispute between the villagers and Green Fuels and the high price of the blended fuel.

Then there is the potential for improved energy access and reduced energy poverty. In order to understand the role of biofuels here, it is important to start at the energy access challenge facing southern Africa.

38 See Moyo et al. (2014).

There are two issues at stake. The first pertains to lack of electricity. Excluding Mauritius and Seychelles, which are small island nations with near 100% electricity access, electricity access rates ranging from a low of 10% in Malawi and 15% in Madagascar (see Table 3). Only South Africa displays a high access rate at 87% (see Table 3). The second pertains to the lack of energy for cooking. The population in many countries in the region, particularly in the rural areas, are heavily reliant on traditional fuels such as wood and charcoal (see Table 3).

In addition, where biofuels are being developed with the intention of creating a large export industry either for raw material or processed fuel, they are unlikely to be available to domestic markets. More importantly, for biofuels to be used domestically there will have to be significant domestic investments. Foreign investors are unlikely to invest in biofuels production in the region without an export mandate, although a minimum quantity could be mandated by countries for the local market.

Increasingly, countries are looking to capitalise on the biofuel opportunity presented at the global stage (see Figure 5). This opportunity arises from policy targets and blending quotas of industrialised countries as well as advanced developing countries such as China to reduce dependence on fossil fuels for transportation, and in doing so reduce their GHG emissions. The European Union's (EU) Renewable Energy Directive of 2009 mandates that 20% of all energy by 2020 should come from renewable energy and the transport sector must use at least 10% renewable energy. Similarly, the United States and Japan have domestic targets for the use of biofuels. Given their own limited availability of land for biofuel development and the abundance of land and favourable tropical climate for biofuels production in Africa, many countries in the EU and Japan are eyeing Africa for the cultivation of biofuels. African policy makers understand this strategic importance of Africa and some even aim to convert this importance into strategic advantage by becoming the primary global suppliers of biofuels.[39]

Questions remain about the extent to which Southern Africa can develop a viable biofuel export market. Biofuel production in Africa is still marginal and large-scale export is non-existent. Investments in crop production, processing facilities and infrastructure are still largely absent. Even though there is heightened appetite from investors, there is still a reluctance to invest due to the high-risk profile of many African countries. At present small amounts of ethanol are shipped from Swaziland and

39 renewables4africa (2006).

Zimbabwe due to preferential access to the EU market, but in general African biofuels are not yet competitive on the world market. Production costs compare unfavourably to those in Brazil. In addition Brazil has embarked on a massive expansion of sugar cane plantations and processing capacity, and will likely take up most of the increase in international trade in ethanol and biodiesel.[40] In the event that a country pursues an export strategy, a balance would have to be found between exports and domestic availability requirements if energy access challenges are to be addressed.

Even when biofuels are available in the domestic market, they cannot be a source of electrification or cooking. They will serve as a fuel for transportation, that too for the urban population since rural transport infrastructure is underdeveloped, in which case they will do little to solve the energy access or energy poverty challenge. Moreover, at current production costs, domestic consumption may not be economically viable in many countries[41] and may have to be subsidised, placing a strain on already stressed fiscal budgets.

What about the potential of biofuels to kick start rural development? The link between biofuels and economic benefits is not as straightforward. Biofuels will contribute to rural development if production and processing is based on small-scale farmers. Wolde-Georis & Glantz point out that the dominant trend to date has been large scale plantations.[42] This in spite of the fact that country's biofuel policies that "contend that small farmers are the main beneficiaries and serve as an opportunity to transition from dependence on traditional biomass fuels".[43] However, how this will happen is not clear. Locally embedded initiatives such as the "development of a small-scale processing industry to enable local farmers to generate fuel from their own feedstock – to provide for the energy needs of rural households – has been slow to emerge".[44] The slow progress on this front has led some to argue that the "nationalist arguments in favour of harnessing natural resources for energy generation to contribute to meeting national energy demand appear to have given way to a reality of corporate refining for external markets".[45] In this regard Mozambique is a case in point (see Box 2).[46]

40 Sielhorst et al. (2008).
41 IEA (2010).
42 Wolde-Georgis & Glantz (2008: 8).
43 (Ibid.).
44 (Ibid.).
45 (Ibid.).
46 Hall (2011).

Box 2: Mozambique Case study[47]

Mozambique has been the frontrunner in embracing biofuels from the 2004 election when the ruling FRELIMO party urged all farmers to plant jatropha on all marginal and unused lands with the objective to turn Mozambique into an oil exporter. Despite poor performance, and evidence that the 'miracle crop' could not withstand harsh agro-ecological conditions, investors who initiated processing facilities for the production of biodiesel from jatropha, and established large-scale sites for cultivation, but also entered into contract farming arrangements. Following directly on the rapid spread of jatropha, and its uneven performance, was the conversion of existing sugarcane production systems to ethanol, and the expansion of sugarcane cultivation to increase supply to processors. After outcries from civil society organisations and a one-year moratorium on new biofuel deals, the government adopted a biofuels policy in 2009, which aims to promote the industry while limiting negative outcomes. None of the implemented projects have met their promised targets for job creation and most have focused on supplying external markets rather than the domestic market.

The experience with biofuels production so far has shown that economies of scale are important in biofuels production, though relatively less important in the production of feedstock than in the processing. More importantly, for countries to become large suppliers of biofuels in the global market and therefore realise benefits of foreign exchange improvement, jobs and higher incomes, both feedstock production and processing will have to be at an industrial scale. On a note of caution, it must be noted that capital-intensive large biofuel farms are also at the mercy of fluctuating global demand and prices.[48]

The threat that large-scale industrial scale production pose to the potential developmental objectives of biofuels has already been highlighted. Industrial scale production is contrary to the small-scale subsistence farming that dominates much of the regional landscape and indeed the continent. It is estimated that almost 70 to 80% of the population in some of the southern African is rural subsistence farmers.[49] As mentioned above, it is possible to promote biofuels through small-scale farmers entering into contrac-

47 (Ibid.).
48 Wolde-Georgis & Glantz (2008).
49 von Maltitz & Brent (2008).

tual supply agreements with refineries. The trade-off here would be the risks to refineries of feedstock availability from multiple small farmers with low yields that may not be adequate to cover the large feedstock demand and higher costs.

The need to achieve scale also raises the possibility of bypassing small-scale subsistence farmers or of consolidating such farmland toward large plantation systems even if it is through leasing of land. This risks not only the displacement of small-scale farmers and poorer land users but also their ability to access basic resources since access to land forms the basis for access to other resources, particularly water. In countries such as South Africa, this may be contradictory to the objectives of land reform that aim to correct historical injustice. Clearly, there could be trade-offs between the ambition to become global suppliers of biofuels and realising benefits for the local and rural economy.

Finally, industrial scale biofuels production is inherently in contradiction with marginal land and will rely on more arable land. This may pose competition for land and water for food crops with risks for food security while generating lower carbon benefits but high secondary biodiversity impacts. Biofuels are already generating a new demand for some agricultural crops, reducing market volatility and surpluses, leading to increases in commodity prices; all of which benefits farmers.[50] High food prices can possibly raise the incomes of farmers when global price movements are transmitted to local markets,[51] implying that households that are sellers of food may benefit through possible increase in income. High food prices also serve as a stimulus for local food production and for the agriculture sector more broadly.[52] Food production is also said to show some increase under most biofuel investment scenarios, especially if biofuel crop interventions address the main reasons for food shortages in rural areas, namely the low productive capacities in these areas.[53]

However, the net impact of biofuels on household food security depends on whether households, specifically low income and poor households, are buyers or sellers of food and their degree of dependence on the agricultural sector for income. In the event that small-scale subsistence farmers and poorer land users are displaced from land in favour of large plantations that are necessary to achieve scale, their ability to withstand

50 Dufey (2007).
51 ODI (2008).
52 (Ibid.).
53 Brent (2014).

food price increases would be minimal. In the absence of appropriate social policies, this would only increase household food insecurity and poverty.

Next is the issue of job creation. Biofuels provide a large number of job opportunities at the local or rural level. For example, it has been estimated that approximately 55,000 jobs could be created in South Africa to meet a 4.5% blending target while over a million jobs could be created in Mozambique if they were to commit 3,379,000 hectares to biofuels production.[54]

However, net job effects would depend on a number of factors. For example, in the event that the biofuels industry is built on small scale farming, there is evidence to show that small-scale agriculture does not create jobs and where it does, the wages are too low to contribute to life improvements.[55] In case of large plantations, the quality of large-scale job creation in feedstock production is under scrutiny with concerns that most jobs in the production of feedstock may largely be manual and poorly paid.[56] This has been seen in Brazil where nearly 80% of the sugar cane harvest is done manually and is regarded as one of the most degrading types of work in the country.[57] These jobs are characterised by low wages, poor working conditions, and prison-like conditions, where heat and malnutrition often lead to worker casualties.[58]

Job creation also depends on where processing takes place. Where biofuels supply chains target export markets, there is the possibility of shipping raw material to importing countries for processing. This may be driven by the need to process the raw material to the standards of the importing countries or the impetus to maintain economic activity and jobs in importing countries. Either way, this implies that there will be no job creation at the final value addition stage.

The above discussion makes it clear that biofuels development has both synergies and trade-offs with the aspirations of African policy makers. The benefits and trade-offs are not straightforward and depend on the strategic decisions taken by countries around their biofuels industry.

54 von Maltitz & Brent (2008).
55 Oya & Sender (2014).
56 von Maltitz & Brent (2008).
57 Rydberg & Cavalett (2010: 131).
58 Barbara (2007).

D. Overview of biofuels policy-making in southern Africa

Biofuel production and consumption is not a recent trend in southern Africa. Blending biofuel into liquid fuels can be traced back to as early as 1920s in South Africa, which continued such blending until the early 1960s when cheap imported fossil fuels made such blending no longer viable and then to the 1980s in Zimbabwe and Malawi.[59] Malawi is the only country besides Brazil that has been blending ethanol continuously on a national basis for more than twenty years.[60] South Africa was the first country in the region to stipulate a formal biofuel policy in 2007 and was followed by Mozambique and Angola. Zambia also has a policy in place, however this policy is not publicly available.[61] Zimbabwe embarked on a process of formulation an integrated biofuel policy in 2015.[62] Table 3 below provides an overview of existing policies.

59 Gasparatos et al. (2012).
60 Sielhorst et al. (2008).
61 Gasparatos et al. (2012).
62 See Bole-Rentel in this publication.

Table 3: Selected country biofuel developments in southern Africa[63]

Country	Biofuels policy	Biofuels mandates/targets	Population with access to electricity in 2012 (%)	Traditional fuel consumption as %age of total energy in 2004
Mozambique	Yes, biofuels strategy	Blend mandate approved but not yet specified	20.2	>80%
South Africa	Yes, biofuels industrial strategy	2% penetration level of biofuels in national liquid fuels pool, or 400 million litres pa. Blending ratio of 2% for biodiesel and 8% for bioethanol	85.4	<20%
Zambia	Yes, biofuels industrial strategy	None	22.1	>80%
Zimbabwe	Draft policy developed	None	40.5	Between 60 and 80%
Mauritius	Ethanol development strategy	None	100	Between 20 and 40%

The policies highlighted above are driven by domestic drivers such as energy security, development of rural economy, and reduced GHG emissions from the transport sector. Lower emissions is a driver only in South Africa which is the highest emitter in the region and is ranked among the top 20 countries measured by absolute carbon dioxide (CO_2) emissions, with emissions per capita in the region of 10 metric tons per annum.[64] However, and as pointed out by Jambe & Mkondiwa "there is a general lack of coherent strategy in Africa to harness its potential" as a hedge against future high oil prices and scarcity, as well as climate change mitigation potential.[65]

63 Table compiled by author based on information from Jambe & Mkondiwa, (2013); The World Bank (2015); and von Maltitz (2013).

64 National Treasury (2013).

65 Jambe & Mkondiwa (2013).

Two issues need to be raised at this point. Contrary to developed countries, sub-Saharan Africa's policies suffer from the lack of supporting strategies that form an integral part of north European and USA policies aimed at stimulating the demand for biofuels processing, the development of infrastructure for processing, expansion of feedstock supplies, development of local markets and investment in research and development.

Failure to develop clear policies and institutional frameworks will inhibit the development of markets that is an "important catalyst for the development of a robust biofuel industry".[66] In the second instance, and in view thereof that biofuel development in Africa should be firmly embedded in rural development rather than serve large commercial interests, any "expansion beyond meeting rural development objectives should be carefully monitored and regulated."[67]

E. Conclusion

As is clear from the discussion above, the future role and impact of biofuels in Africa are unclear. Apart from environmental and social concerns, there are also issues relating to markets and uptake of biofuels – whether there is enough demand internally, firstly, and secondly, whether Africa can be competitive with existing biofuel markets.[68]

African policy makers are eyeing biofuels to invigorate the rural economy, reduce energy poverty and exploit export opportunities. It is in this spirit that many southern African countries have formulated or considering strategies for biofuels development. Policy-making must recognise biofuel is not a homogenous product. Even the same product may have a wide range of environmental outcomes.

Moreover, as this paper cautions, there is every possibility that biofuels may not deliver on these aspirations, leading instead to unintended adverse environmental and social consequences. For biofuels to realise the envisioned potential, maximise benefits and minimise risks, they need to be carefully rooted in the country context, managed prudently for environmental sustainability and social impacts.

Any decision regarding the development of biofuels development requires country-specific assessments to ascertain the viability and econom-

66 (Ibid.).
67 (Ibid.).
68 Sielhorst et al. (2008).

ic feasibility of biofuels. More importantly, government should 'define the rules of the game' to ensure that rural communities are protected from losing their land, the equitable division of the spoils between rural communities and investors and to prevent biofuels being grown in protected areas. Aligned to this the development of national markets to support the industry.[69]

To conclude, biofuel development should be part of a larger sustainable development countrywide strategy that also include other renewable energy options, and speaks to important goals such as the Sustainable Development Goals. Arguably such an integrated approach will unlock complementarities that can maximise benefits and minimise adverse outcomes.

Bibliography

Amigun, B *Sustainability of biofuels in Africa* (2011) keynote presentation at the Biofuel International Conference, 14 to 16 November 2011 in Ouagadougou, Burkina Faso.

Baier, S, M Clements, C Griffiths & J Ihrig, 2009, *Biofuels impact on crop and food prices: Using an interactive spreadsheet*, International Finance Discussion Papers Number 967, at http://www.federalreserve.gov/pubs/ifdp/2009/967/ifdp967.pdf, accessed 18 July 2016.

Barbara, JS, *The false promises of biofuels*, 2007, The International Forum on Globalization and the Institution of Policy Studies, at http://www.ifg.org/pdf/biofuels.pdf, accessed 20 July 2016.

BirdLife International, European Environmental Bureau, Fern, Friends of the Earth, Oxfam and Transport & Environment, 2009, *Biofuels, handle with care. An analysis of EU biofuel policy with recommendations for action*, at https://www.transportenvironment.org/sites/te/files/media/2009%2011_biofuels_handle_with_care.pdf, accessed 20 July 2016.

Brent, A, 2014 *The agricultural sector as a biofuels producer in South Africa. Understanding the Food Energy Water Nexus*, 2014, at 6__a16270_the_agricultural_sector_as_a_biofuels_producer_in_sa___online.pdf, accessed 18 July 2016.

De Fraiture, C, Y Liao & M Giordano, 2008, Biofuels and agricultural water use: Blue impacts of green energy, *Water Policy* 10(S1), 67.

Department of Minerals and Energy, Republic of South Africa, 2007, *Biofuels Industrial Strategy of the Republic of South Africa.*

Dufey, A, 2007, *International trade in biofuels: Good for development? And good for environment?* International Institute for Environment and Development (IIED).

Ewing, M & S Msangi, 2009, Biofuel production in developing countries: Assessing tradeoffs in welfare and food security, *Environmental Science & Policy* 12, 520.

Fargione, J, J Hill, D Tilman, S Polasky & P Hawthorne, 2008, Land clearing and the biofuel carbon debt, *Science* 319, 1235.

69 Jambe & Mkondiwa (2013).

FAO / Food and Agriculture Organization of the United Nations, 2008, *The state of food and agriculture*, at ftp://ftp.fao.org/docrep/fao/011/i0100e/i0100e.pdf, accessed 18 July 2016.

FAO / Food and Agriculture Organization of the United Nations, 2011, *Bioenergy environmental impact analysis (BIAS)-conceptual framework*, prepared by Oeko-Institut, IFEU and Copernicus Institute for FAO.

Gasparatos, A, LY Lee, GP von Maltitz, MV Mathai, JA Puppim de Oliveira & KJ Willis, 2012, *Biofuels in Africa. Impacts on ecosystem services, biodiversity and human well-being*, at https://www.cbd.int/cop/cop-11/doc/vtable/Biofuels_in_Africa.pdf, accessed 18 July 2016.

Gerbens-Leenes, PW, AY Hoekstra & T van der Meer, 2009, The water footprint of energy from biomass: A quantitative assessment and consequences of an increasing share of bio-energy in energy supply, *Ecological Economics* 68, 1052.

Hall, R, 2011, *Investor rush in Southern Africa: Towards a typology of commercial land deals*, ICAS Review Paper Series No. 2, at https://www.tni.org/files/Hall%20 ICAS%20WP%202.pdf, accessed 19 February 2016.

IEA / International Energy Agency, 2010, *Sustainable production of second-generation biofuels. Potential and perspectives in major economies and developing countries.*

Jacobson, MZ, 2007, Effects of ethanol (E85) versus gasoline vehicles on cancer and mortality in the United States, *Environmental Science & Technology* 41(11), 4150.

Jambe, CBL & M Mkondiwa, 2013, Comparative analysis of biofuels policy development in Sub-Saharan Africa: The place of private and public sectors, *Renewable Energy* 30, 614.

Kunen, E & J Chalmers, 2010, *Sustainable Biofuel Development Policies, Programs, and Practices in APEC Economies. Project No. EWG 19/2009 under Asia Pacific Economic Cooperation's Biofuels Task Force*, at http://www10.iadb.org/intal/intalcdi/PE/2010/07493.pdf, accessed 25 January 2016.

Moyo, P, M Moyo, D Dube, O Rusinga, 2014, Biofuel policy as a key driver for sustainable development in the biofuel sector: The missing ingredient in Zimbabwe's biofuel pursuit, *Modern Applied Science* 8, 1.

Nadia, O, 2009, *Bioenergy for Africa: An illusion or a sustainable option to reduce the vulnerability to energy and poverty?*, Presentation at the 32nd IAEE International Conference, San Francisco, 21 to 24 June, at http://www.usaee.org/usaee2009/submissions/Presentations/SFNadia.pdf, accessed 22 February 2016.

National Treasury, Republic of South Africa, 2013, *Policy paper for public comment. Carbon tax policy paper: Reducing greenhouse gas emissions and facilitating the transition to a green economy.*

ODI / Overseas Development Institute, 2008, *Rising food prices: A global crisis. Action needed now to avert poverty and hunger*, Briefing Paper.

Oya, C & J Sender, 2014, *Agriculture, economic development (and the issue of land reform)*, Presentation and lecture at the African Programme on Rethinking Development Economics.

Reinhardt, G, K Becker, DR Chaudhary, J Chikara, E von Falkenstein, G Francis, SO Gärtner, MR Gandhi, A Ghosh, PK Ghosh, HPS Makkar, J Münch, JS Patolia, MP Reddy, N Rettenmeier & SC Upadhyay, 2008, *Basic data for jatropha production and use*, at http://jatropha.pro/PDF%20bestanden/en_jatropha-database-june-2008[2].pdf, accessed 20 July 2016.

renewables4africa, 2006, *African nations form biofuel group*, at http://energy4africa.net/news/viewnews.php?ID=173, accessed 2 February 2016.

Rosegrant, MW, 2008, *Biofuels and grain prices: Impacts and policy responses. Testimony for the U.S. Senate Committee on Homeland Security and Governmental Affairs*, at http://large.stanford.edu/courses/2011/ph240/chan1/docs/rosegrant20080507.pdf, accessed 20 July 2016.

Rydberg, T & O Cavalett, 2010, "Lessons and reflections on Brazilian ethanol developments: What can Africa learn?" in: Nordiska Afrikansitutet (ed), *Biofuels, land out sourcing and food security in Africa*, 131.

Sielhorst, S, JW Molenaar & D Offermans, 2008, *Biofuels in Africa: An assessment of risks and benefits for African wetlands,* at https://www.unido.org/fileadmin/user_media/UNIDO_Header_Site/Subsites/Green_Industry_Asia_Conference__Maanila_/GC13/Wetlands.pdf, accessed 22 February 2016.

Sulle, E, 2013, *Biofuels investments in Tanzania: Policy options for sustainable business models*, at http://www.plaas.org.za/sites/default/files/publications-pdf/Sulle%20PLAAS%20Seminar%202013_%20Biofuels%20investment%20in%20Tanzania_March%2014_Final%20(2).pdf, accessed 19 February 2016.

The World Bank, 2015, *World Development Indicators: Electricity production, sources, and access*, at wdi.wordbank.org/table/3.7, accessed 27 February 2016.

Timilsina, GR & A Shrestha, 2010, *Biofuels: Markets, targets and impacts*, Policy Research Working Paper 5364.

UNEP / United Nations Environment Programme, 2009, *Towards sustainable production and use of resources: Assessing biofuels*, at http://www.unep.org/pdf/biofuels/Assessing_Biofuels_Full_Report.pdf, accessed 15 January 2016.

von Maltitz, GP & A Brent, 2008, *Assessing the biofuel options for southern Africa*, Paper presented at the 2nd CSIR Biennial Conference 17 to 18 November in Pretoria, South Africa.

von Maltitz, G, 2013, *The role of biomass energy in Southern Africa*, Presentation at a conference on best practice in the governance of Africa's dry land forests: Implications for Southern Africa 22 October, at http://www.saiia.org.za/doc_view/392-drylands-conf-s3-vonmaltitz, accessed 27 February 2016.

Wang, M, M Wu & H Hong, 2007, Life-cycle energy and greenhouse gas emission impacts of different corn ethanol plant types, *Environmental Research Letter* 2, 024001.

Wolde-Georgis, T & MH Glantz, 2008, *Biofuels in Africa: A pathway to development*, at http://www.ccb-boulder.org/wp-content/uploads/2013/01/biofuels-in-africa.pdf, accessed 20 February 2016.

PART III:
BIOFUELS REGULATION –
A COMPARATIVE LAW PERSPECTIVE

Industrial dynamics and internationalisation of Brazilian ethanol supply chains – a brief historical review on regulation and its impact

Prof Dr Renata A Marcheti, Universidade de São Paulo, Faculdade de Economia, Administração e Contabilidade de Ribeirão Preto, Brazil

Prof Dr Antonio D Padula, Universidade Federal do Rio Grande do Sul – UFRGS, Escola de Administração and Centro de Estudos e Pesquisas em Agronegócios[1], Brazil

A. Introduction

The objectives of this study are concerned with the regulatory framework as instrument of incentive or disincentive for ethanol production in Brazil, the expansion of ethanol production, the process of concentration, and the internationalisation of the ethanol supply chains in the state of São Paulo. The ethanol industry is beginning to take on clearly oligopolistic characteristics: the Concentration Index CR_8 varied from 32.03% to 36.89% (2006 to 2009). Brazilian exports gained relevance and leapt from 1% of total production in 2000 to 17% in 2008. In 2009, the share of foreign investment in the Brazilian ethanol sector achieved approximately 16% of the sector's assets. The industry faces the great challenge of turning ethanol into an international commodity.

Society and the law or regulation are, mutually and continually, influencing each other. Society is constantly transformed by law and regulations and, at the same time, law and regulations are transformed by society. Within the legal field, law and regulation have different meanings. Statute, common law and other terms are also considered to be different means of control. By 'law or regulation' for the purpose of this article we mean the rule of conduct or action formally prescribed and formally recognised as binding or enforced by a controlling authority, usually subordi-

---

1 With the collaboration of Mariana Boeira, Suellen Moreira de Oliveira, Armando Fornazier, Jorge Estuardo Tello Gamarra, Moises Centenaro, Raoni Fonseca Duarte.

nating, systematised in a legal system, and capable of governing people's social lives. Brazilians live under a very complex legal system, based on civil law tradition. Despite this complexity in law, Brazil has become one of the largest producers of biofuels such as ethanol and biodiesel, and mainly exactly because of the public policies driven by laws and regulations. As has been shown in previous studies of industrial organisation, there is a long-term tendency in many industries towards concentration, resulting in an ever-smaller number of firms. Another phenomenon found within the industrial dynamic is the process of internationalisation of firms as a means of expanding their markets and increasing their productive capacity. Among the many questions related to the internationalisation of the ethanol industry, one of the main questions is whether Brazil will be a global player or only a manufacturing platform for raw materials in this global value chain. Therefore, the problem statement and the objectives of the present research are concerned with the expansion of ethanol production, the behaviour of the internal market, and whether the necessity for gains in scale are influencing the process of concentration and internationalisation of the ethanol supply chains in the state of São Paulo, which is responsible for approximately 60% of Brazilian ethanol production. The analytical framework underpinning the research explores and integrates two conceptual approaches: a) industrial organisation (the structure-conduct-performance paradigm, and the processes of mergers and acquisitions); and b) internationalisation of the firm (behavioural and economic theories, market power theory, eclectic paradigm, and global value chain). The research methods and procedures occurred in two steps. In a first, quantitative step, based on data obtained from databases published by the most important institutions within the sugar/alcohol sector: the Sugarcane Industry Association (*União da Indústria de Cana-de-Açúcar – ÚNICA*), the Bioenergy Producers Association (*União dos Produtores de Bioenergia – UDOP)* and the Ministry of Agriculture (*MAPA*), a dynamic of structural change was identified in the sector, based on the rate of concentration (CR_K) proposed by Bain[2] followed by the Herfindahl-Hirschman (HH) index. Another stage was the literature research, especially into journal sources and industry yearbooks in order to monitor the evolution of the sector. In the second, exploratory step, three specialists from representative entities of the alcohol sector and executives from two distilleries producing ethanol were interviewed in order to identify the strategies and processes of internationalisation within the ethanol sector.

---

2 Bain (1954 and 1968).

B. A brief retrospective

Sugar from sugarcane is an ancient product for human consumption but its market only expanded during the Age of Discovery. It was brought to America in 1493 by Christopher Columbus, and the first sugarcane mill in the Portuguese colony – Brazil – was built in 1532 by coloniser Martim Afonso in São Vicente, in the vicinity of present-day Santos, a city in the state of São Paulo. Although the sugarcane plantations and the mill were initially established in São Paulo, it is known that small plantations and mills proliferated in many other Brazilian captaincies, especially in the northeast, supported by Portugal. Only a small part of the sugar produced in Brazil by then was destined for the production of byproducts.

To secure its strategic position, Portugal needed to colonise and exploit the northeast coastline rapidly. It was decided to cultivate sugarcane because it was easy to plant and harvest this fast-growing crop and it was well-adapted to the climate and soil of the region. Besides that, it would be easy to channel surplus sugar production to Europe where the consumption market was inflating its price. So Portugal incentivised the beneficiaries of land to produce sugar, using primarily slave labour. The incentive was aligned with the most important value at the time, namely economic and political status. From a social point of view this was the first 'policy' in respect of an ethanol incentive in Brazil. By then, one of the by-products from sugar was alcohol from which a spirit called *aguardente* (literally translated as 'burning water') was, and still is, produced. *Aguardente* is the basis of Brazil's most famous drink – the *caipirinha* – and the historic incentive for the production of sugar thus led to the bigger artisanal production of alcohol and *aguardente*.

By the end of the 16th century, with the help of Portugal's incentives, the Pernambuco area became the leading sugar-producing area in the country, with over 70 plants; followed by the Bahia captaincy. Brazil had already become a big player in the production of sugar. The colonisers had made huge profits, selling the colony's production in Europe, which had been enriched by receiving the gold and silver extracted from American soil. With its wealth, Europeans consumed more sugar and the producers in Brazil rapidly prospered.

In the mid-17th century, the second planned governmental intervention in sugar production occurred – this time by way of direct regulation. Sugar production was undergoing a serious crisis owing to the political situation in Europe, and the Portuguese king and Brazilian emperor, Dom Pedro II, with the support of his ministry, set in place the so-called Pragmatic Laws.

This regulation forbad the use of certain foreign products in order to reduce imports and diminish the Luso trade deficit; and reoriented the productive activities in the kingdom and in the colonies, with the help of foreign technicians.

The reorientation of the industry included incentives to boost manufacturing through coordinating activities, organising the supply of raw materials, standardising sales and tax rates, and creating measures to facilitate the marketing of products, etc. Besides that, the Crown forbad the readjustment of the selling price of imported products in order to reduce the profit margin of the dealers and decrease the export of currencies. This became set national economic policy, characteristic of the late 17th century and much of the 18th century.

The approach to the sugar industry was to insert new technologies and modernise the activity. The modernisation of sugar production in Brazil came with investments and partnerships between investors and landowners, where the investors would take care of the industrial process (central mills) and the landowners would cultivate the land.

This model of production was not successful, however, owing to the political and economic structure which was in place, but the capital invested helped the industrial process to evolve, and it was therefore considered an incentive. Besides, it was the precursor to some recent production models where landowners and industry owners contract production under what is called agricultural partnership contracts – with land-leasing and sharecropping.

In the 1800s, coffee was becoming an important product and by the end of the 19th century, 70% of the world coffee production came from Brazilian farms, concentrated especially in the countryside of São Paulo State.

From 1822 immigrants began arriving in Brazil when the coffee boom was in full swing. Over 4 million people came to the country between 1822 and 1932 – especially after 1889, mostly because slavery had been abolished through the Golden Law decreed in 1888. The coffee farms needed workers. The conditions for immigrants were not perfect, but the situation in Europe was critical at that time. Brazil had cheap land to sell to the immigrants – though not suited for coffee crops. Labour laws were for the most part fair and the immigrants' reception in the country was hospitable.

Europeans who came to Brazil at the time were mainly Italians who bought pieces of land, started their own alcohol production, while working on the coffee farms concentrated in the São Paulo State.

This immigration promoted the installation of uncountable mills especially in the regions of Campinas, Itu, Moji-Guaçú and Piracicaba. Later, new mills were established in the Ribeirão Preto region. This huge immigration movement raised the need for sugar, not only because more people were living in the country, but because coffee was prepared for consumption with sugar. The sugar produced in Brazil was then directed to the internal market and this kept the sugarcane industry alive in a period when beetroot sugar dominated most of the external market.

During World War I the beetroot fields in Europe were devastated and in 1918 beetroot sugar represented only 25% of the world's total consumption. This presented a good opportunity for the Brazilian sugarcane industry. In the early 1900s, even the *aguardente* producers turned to the production of sugar, because the demand and the price were high.

With the invention of Ford's Model T car in 1908, which could run on gasoline, kerosene or ethanol, the sugar plants – that were once alcohol mills – saw an added advantage in the opportunity, but it was not before 1925 that the first experience with combustible ethanol took place in Brazil.

In 1931 the first ethanol production incentive law was passed in Brazil. Federal Law 19.717/31, effective in the whole country, set in place a mandatory purchase of ethanol by gasoline importers for blending as a gasoline additive. Up to 5% of the total imported and consumed gasoline in the country was supposed to be substituted by ethanol. This was the first move towards a public regulatory policy framework on ethanol in Brazil.

In 1933, through Decree 22.789, the federal government of Brazil created the Sugar and Alcohol Institute (IAA), a regulatory organisation that had as its objective the orientation, stimulation and control of the production of sugar and ethanol – and its raw material – in all national territory.

The newborn institute consolidated the federal interventionist policy in the sugar and alcohol industry in the country.

Before the inception of the IAA, federal intervention was implemented on an emergency basis only. Mill owners favoured the introduction of an institute to regulate sugar and alcohol production, especially because they were in the middle of an overproduction crisis in the sector – the production of sugar was exceeding the domestic consumption and the same situation was seen in other countries which prevented exportation of surpluses. The production of alcohol needed government support and the producers saw the creation of a local authority reporting directly to the president as an important measure.

A curious fact is that when in 1933 the internationally known Rio de Janeiro Grand Prix was created, gasoline supplies were so low that people wondered how long the race event would survive. Brazil's National Petroleum Council even pleaded for postponement of the event. IAA decided to supply the fuel, and a blend of 90% ethanol and 10% benzene was used in the races. It was not a regulation, but it certainly was an incentive to the ethanol sector.

IAA effectively set in place annual plans to protect the sugar, alcohol and *aguardente* industry, through the establishment of quotas for distilleries and sugarcane suppliers – disincentives for those with bigger production capacity and incentives for those with smaller capacity.

World War II broke out and made it difficult to bring sugar to the central and southern parts of Brazil. So through a completely different regulation, IAA incentivised the creation of new mills in those regions from 1940 onwards. The purpose of the IAA was to raise the production of sugar in the country as much as possible. Thus the industry grew.

In the first years after the war, exportation was the main route for the overproduction raised by Law Decree 9.827/46. But the market was too small for all that sugar, and prices came down. In 1948 the Brazilian federal government decreed measures under n. 25.174-A to stimulate alcohol production as car fuel.

It is important to note that at that time there was an economic deficit created by the crescent importation of gasoline, which drew out the cambial reserves of the country. Blending diminished that negative result, but industry was still more dedicated to sugar than to alcohol. Interestingly, it was the first time that the producers were obliged to apply part of the price per 60 kg bag of sugar to the industrial and agricultural workers, as pharmaceutical/medical and social benefits. This became regulation.

Production grew as the population and the internal demand for sugar grew. But when modern technologies and modern business administration techniques were used and promoted, sugar production soon generated surplusses and industry was pressured to go to the international market.

In the 1950s, the sugar agroindustry expanded significantly since the cities were evolving and the central and southern parts of the country were flourishing. The country began producing much more than it needed and exportation was becoming an imperative. Brazil was again one of the biggest exporters of sugar, all promoted by IAA actions and acts. With many resolutions coming into effect to regulate and incentivise the sector, Brazil was producing more than it needed and exportation was the only way out, although the product would have to go through the so-called Wide Free World Market, which was already becoming saturated.

IAA needed to seek new shares in the international market and expand the internal market.

In 1959 the Cooperative for Sugarcane, Sugar and Alcohol Producers (COPERSUCAR) was created in the state of São Paulo as a private association having the purpose of commercialising production on behalf of its members. It was the first Brazilian organised marketing company for sugar and ethanol, though export would necessarily go through IAA.

In 1960 IAA was moved and became part of the Ministry of Industry and Commerce, but it continued its efforts towards maximisation of exports and to increase the productive capacity of the existing crops and mills.

In 1961 the export division within IAA was created, and the regulation for international transactions was promulgated through Decree 50.818/61.

Through Decree 51.104/61, the federal government created the Fund to Recover the Sugarcane Agro Industry. This was a meaningful incentive for the sector. In the same year another fund was created to stimulate and consolidate the sales of sugar internationally, namely the Special Fund for Exportation of sugarcane products. IAA was, by law, the only exporter of sugarcane products.

In 1962 IAA signed a contract with the Bank of Brazil to obtain financial resources to be applied in projects designated to improve the use of land in the sugarcane production zones, so that producers would not lack the basic material for the production of sugar.

In 1963 IAA created a programme that later became the Plan for the Expansion of the National Sugar Industry (Resolutions 1761 and 1762). It is important to mention that this measure was the first legally implemented action to create a regulatory stock of sugar in the country.

The policy to maximise the sugar/alcohol production in the country and facilitate exports was in place and ongoing from the 1960s and 70s through IAA measures. At the end of the 1960s IAA started a policy to concentrate and modernise the structure of the Brazilian sugarcane agroindustry to make it more competitive in the international market.

In 1968 IAA initiated the development of a strategy implemented in the country and production became competitive in terms of price and quality. This reflected in the Special Fund for Exportation since the prices paid to the producers was lower than the average price IAA received from the exports.

Concomitantly, the private sector through COPERSUCAR established its Copersucar Technology Center in 1969, to develop technologies to improve production of sugarcane and its sub-products.

In 1971 IAA set in place the National Program for Sugar Cane Improvement (PLANALSUCAR) and the Rationalization Program of the Sugar Agribusiness.

The purpose of PLANALSUCAR was to conduct research for the improvement of the quality of sugarcane crops. The programme was so important that it still exists, with a completely different composition but with the same purpose, and many high-tech university centres form part of it today. It was the seed to make Brazil one of the most advanced research hubs in the world as far as sugarcane is concerned.

The Rationalization of the Sugar Agro-Industry (Decree 1186/71) transformed into the Program of Support to the Sugarcane Agroindustry (Decree 1266/73) and had its objectives linked to the better dimensioning and use of the installed industry capacity, to improve the efficiency of the productive system, and to structure the commercialisation system from the agro to industry aspect.

To improve the quality of crops and rationalise production, IAA financed mergers and acquisitions of sugarcane agroindustries and the construction of structures to facilitate exportation (like the establishment and operation of sugar terminals in two major ports of the country), through the Special Fund for Exportation. All the IAA programmes together were really a 'put in place policy' for the advancement and growth of the sugar and ethanol industry and was quite successful in achieving Brazil's goals laid out in its Plan for the Expansion of the National Sugar Industry drawn up in 1963.

It is noteworthy that in the 1960s the United Nations was concerned with environmental issues and many international treaties were signed under its influence. In 1972, in Stockholm, the United Nations Environment Programme (UNEP) was created during the International Conference on the Environment.

Many laws, resolutions and decrees were issued during this period in Brazil, all giving the industries better conditions to produce and sell, and through Law Decree 1266/73 IAA activated even more of the activities of the Fund, financing mills to better implement the structure and raise production. Environmental issues were, however, not put aside.

In 1973 the Brazilian government created the Secretariat for the Environment as part of the Ministry of the Interior.

The Secretariat was created right in the middle of a period where the Ministry of Industry and Commerce, through IAA, was launching major projects to accelerate the growth of a sector that, of course, had environmental impacts.

Many businesses took loans from the Fund, but some of them did not make the correct investments. On top of that, in 1979, the prices of sugar in the external markets started to drop abruptly, exportation was not a good option anymore, the resources of the Fund diminished, and many producers could not make enough to pay their debt to the federal government.

In 1973 the oil crisis struck. OPEC (Organization of the Petroleum Exporting Countries) increased prices to protest against the United States position in the Yom Kippur War. By 1974 the price of oil had risen 300% and it impacted on all economic transactions, including exportation of sugar. The rapid growth of the Brazilian sugarcane industry had become history by then. It was the end of the so-called Brazilian Miracle. And the Brazilian government had to act fast to recover the industry and provide a destination for the sugar produced and not exported.

In 1975, the PROALCOOL (National Alcohol Program) was introduced by Law 76593/75. Like the name says, it was a nationwide incentive programme through which the federal government financed the substitution initiatives of petroleum-based fuels by ethanol, specifically anhydrous ethanol, used as an additive of gasoline, substituting the tetraethyllead lead. The sector glimpsed a chance of immediate recovery.

The government boosted ethanol production by offering loans and subsidies to the energy sector, determined that Petrobras would buy a minimum annual volume of ethanol, as well as transport, distribute, and blend it with gasoline. The government would set the price of ethanol through the Instituto do Acucar e Alcool (IAA).

Decree n. 80.762/77 (adopted in 1977) superseded the 1975 decree and established parity between ethanol and sugar prices. During the following 10 years, US$16 billion was invested in genetic research to improve sugarcane yield, subsidise the ethanol sector, and underwrite low-interest financing for new agricultural machinery.

In September 1979, ANFAVEA (Brazil National Association of Motor Vehicle Manufacturers) and the federal government signed an agreement to a two-way collaboration – where the manufacturers were to pursue new technologies to produce ethanol vehicles and the government would incentivise the consumption of those vehicles – and of ethanol.

The Brazilian government wanted to eliminate its dependence on imported oil. The Iranian Revolution had started and a general panic about the rapidly decreased oil supply took hold, elevating the price of the barrel even more. Of all the petroleum used at the time, 80% came from overseas.

Prototypes were developed by ANFAVEA members, especially Fiat, Volkswagen, GM and Ford. In May 1979 there were 16 gasoline stations supplying hydrous ethanol for a fleet of 2,000 neat ethanol-'adapted' vehicles.

In July, the Fiat 147 was launched into the market, becoming the first commercial, new technology, neat ethanol-powered car sold in the world. Thereafter, all Brazilian carmakers modified gasoline engines to support hydrous ethanol characteristics.

The federal tax on industrialised products (IPI) imposed on those ethanol-fuelled vehicles was reduced. The idea was to overcome consumer reluctance and distrust of the new car and fuel. Ethanol-fuelled taxis were granted a tax exemption.

By the end of 1979, Brazil had a neat-ethanol fleet of 4,614 cars corresponding to 5% of the total car production in the country. Six years later (1985) it was almost 75%.

In 1986 ethanol suffered a downturn. Brazil faced an economic crisis, sales dropped drastically and inflation went up.

In 1989 inflation soared – reaching 1,764.86% in January.

The oil price was going down. Inflation control became imperative since Brazil had just entered a democratic period. Brazil's dependence on foreign oil decreased as the country began relying more on domestic petroleum supplies. Government started to remove the incentives for ethanol in order to equalise balance of trade. Competitiveness suffered, leading to production cuts, and the ethanol produced was not enough for the consumption. The programme became discredited, while high international sugar prices and lower government incentives spurred the ethanol industry to turn to sugar production. The fact was that 90% of vehicles traded were ethanol-only, and there was no ethanol to buy.

The mandatory blend of ethanol fuel with gasoline continued to be used and it fluctuated between 10% and 22%, from 1976 until 1992.

In 1990, the ethanol sector was deregulated and the once-powerful IAA was terminated.

In 1993 a law established a mandatory blend of 22% anhydrous ethanol in the entire country. This same statute permitted the executive to alter the percentages of ethanol depending on the national scenario of ethanol production, but within pre-established boundaries.

This was not enough.

By 1994, only 12.2% of motor vehicles sold in Brazil were ethanol-only. This downturn reflected a lack of consumer confidence, which stemmed from hardships faced by owners of ethanol-only vehicles during the supply crisis. In 1998, ethanol-only vehicle production was discontinued.

The industry had to be modernised in order to maintain the competitiveness of Brazilian ethanol in internal and external markets. Agriculture was mechanised, scientists experimented with new variations of sugarcane, strategic mergers and acquisitions were carried out, and mills and distilleries were modernised.

In 1997 the federal government created a new agency, the National Petroleum Agency (Agência Nacional do Petróleo (ANP)), to regulate and monitor the exploitation, production, refinement, transportation, distribution, retail, imports and exports of oil, derivative substances, natural gas, and ethanol.

Finally, in 1999, ethanol prices became subject to free market rules and were no longer set by the government.

Proalcool and the Brazilian ethanol experience had failed. However, a new rise in international oil prices, and increasing international awareness about climate change, as well as the development of flex-fuel engines in Brazil, gave Brazilian ethanol a new push. The spike in oil prices caused by instability in the Middle East revived the discussion about fossil fuel substitutes and alternative energy sources.

The 1997 Kyoto Protocol also called attention to alternative energy sources, specifically regarding the need to reduce greenhouse gas emissions. However, it was the development in 2003 of a commercially viable flex-fuel engine that could run on gasoline, hydrated ethanol, or any blend of gasoline and anhydrous ethanol that truly propelled ethanol back into dominance.

The Brazilian flex-fuel engine runs on any combination of anhydrous ethanol and gasoline blend, as well as on pure hydrated ethanol. In contrast, flex-fuel vehicles produced in other countries operate on a blend of a maximum of 85% ethanol.

Brazil's flex technology dissipated fears about the unavailability and price of ethanol and gave Brazilian drivers the ability to fuel their vehicles with either ethanol, gasoline, or both. For these reasons, the flex-fuel vehicle was a big success, significantly increasing the national demand for ethanol, with half of the 30,000 gas stations in the country offering hydrated ethanol. Blending was still in place, though. In 2003 these limits were fixed at a maximum of 25% and a minimum of 20% of ethanol by volume.

By 2006, seven in every 10 vehicles sold in Brazil had flex-fuel engines. In 2007, the ethanol limit went up to 25%, and in 2010 the government determined a reduction of the percentage from 25% to 20% for 90 days, owing to the supply shortage and the high price of ethanol. In the same year it went up again to 25%. In the harvest season of 2010/2011, Brazilians experienced another shortage and the mandatory blend was set at 18% of ethanol, in April. It was a difficult year and Brazil had to import ethanol from the United States, again. President Rousseff's policy to fight inflation, since her first mandate, by capping gasoline costs at the pump was also limiting prices for sugar-based ethanol, discouraging millers from investing in new output and exacerbating the problem.

Some change was expected after her second election since she had negotiated extensively with the sector and had made promises. Nothing improved, however. Petroleo Brasileiro SA, the state-run oil company, was boosting money-losing gasoline imports at its fastest pace ever, holding down prices for alternative fuels and hindering investments in ethanol.

The economics were prompting some sugarcane companies to close their facilities. About 14 mills were shut in the 2012–13 harvest season, following 16 closures in the prior season.

In March 2015, the government again raised the ethanol/gasoline blend boundaries from 25% to 27%, except on premium gasoline that was kept at 25%, owing to a movement of ANFAVEA that justified the need to maintain one option of the 25% maximum to serve the consumers that owned cars developed to run on a maximum blend of E25.

By April of 2015, 60 mills had been closed and almost 80 were declaring bankruptcy. At present, there is no legal framework specifically aimed at regulating ethanol in Brazil.

Production and consumption is governed primarily by the National Energy Policy (Law n. 9.478/97), under its general regulation of fuels. In the current regulatory landscape, Agencia NacPetroleo (ANP), the National Petroleum Agency, is the main regulatory agency for any activity regarding biofuels. The Ministry of Agriculture, Livestock and Provisions (Ministério da Agricultura, Pecuária e Abastecimento – MAPA) and the Interministerial Commission of Sugar and Ethanol (Comissao Interministerial do Açúcar e do Alcool, or CIMA) play strategic roles in defining the percentage of the mandatory blend of ethanol to gasoline on federal land.

C. Market concentration

Why some companies outdo others or why some show improved performance is a major question investigated within the fields of strategic management and industrial economics.[3] Researchers have started to answer this intriguing question via different paths. Some claim that the source of advantage comes from resources,[4] while others maintain that it comes from the relationships existing among companies.[5] There are also scholars who defend the theory that superior performance comes from minimising transaction costs.[6] However, there are academics who argue that the market framework in which a company operates is the key vector for an organisation's possibilities of achieving superior performance. The last approach is known as the structure-conduct-performance paradigm.

Under the last paradigm, companies operate along a 'continuum of market structures' that ranges from perfect competition (zero concentration) all the way to monopoly (total concentration – one single supplier). Hence, the goal of companies inside an industry lies in seeking concentration on a few suppliers. Once that concentration is achieved, a company's better performance in the industry will be a natural consequence.

More specifically, through the structure-conduct-performance paradigm, Bain[7] sought to demonstrate the hypothesis that the average profit of companies operating in concentrated markets is significantly higher than that of companies operating in less concentrated markets or industries with atomised structures. Therefore, Bain started by defining industry as a group of companies that manufacture the same product or perfect substitute products. In turn, the degree of industry concentration is defined as the combined production volume rate of a given group of companies (one, four, eight or twenty) against the total volume manufactured by the entire industry.

It should be noted that Bain's study[8] also highlights the importance of entry barriers as a structural dimension that influences company performance in a given market. In the study, the author ranks entry barriers as: (1) very high; (2) substantial; and (3) moderate to low. From that standpoint, in the short and medium terms, companies that have managed to put

---

3 Rumelt et al. (1991); Hölmstrom (1998); Barney (1991).
4 Penrose (1959); Wernerfelt (1984); Grant (1991); Barney (1991); Peteraf (1993).
5 Dyer & Singh (1998); Gulati et al. (2000); Anand & Khanna (2000).
6 Coase (1937); Williamson (1975, 1985, 2002 and 2010).
7 Bain (1951 and 1954).
8 Bain (1954).

up 'higher' entry barriers will be more likely to keep up superior performance.

To Bain,[9] the condition of entry into an industry is dictated by the advantage in terms of cost or prices that established companies have over potential entry candidates, and can be measured by the degree to which established companies are able to raise their prices above minimum competitive values – however, without making the entry attractive to other competitors.[10]

With respect to a company's conduct, the aforementioned emphasises the behaviour pattern companies adopt to retain or increase market concentration. To do that, a company's conduct is directed mainly to: (a) pricing policies and strategies; and (b) processes or mechanisms to interact and coordinate competition policies among market competitors.[11]

In turn, performance is related to the aggregate results companies reach through their conduct standards, so that their main dimensions of analysis are: (a) relative technical-productive efficiency; (b) price level related to long-term marginal production costs and average long-term marginal production costs, and to the resulting profit margin; (c) actual output of the industry related to the maximum possible quantity consistent with price equality and long-term marginal costs; (d) investments in sales and promotion in relation to production costs; (e) characteristics of the product or products, including design, quality level, and variety; and (f) the industry's high rate of progress in developing both products and production techniques.[12]

As we have seen so far, besides the structure-conduct-performance paradigm, there are a variety of currents that also try to explain higher company performance. Nevertheless, after Bain,[13] other researchers[14] continue finding evidence that such a paradigm still accounts for the market's structural dynamics as well as for superior company performance.

---

9 Bain (1968).
10 (Ibid.).
11 Bain (1968); Ferguson & Ferguson (1994).
12 Bain (1968).
13 Bain (1951, 1954 and 1968).
14 Qualls (1972); Hou & Robinson (2006).

D. *Internationalisation*

Some theoretical approaches have been developed to understand the firm's internationalisation. The *theory of the growth of the firm* shows that companies tend to accumulate capital and grow. Firm growth depends on the opportunities available. Such opportunities may lead the firm to step into other markets besides the company's original market, and the international market can be an option. It should be noted that the company's current market growth rate determines the opportunities.[15]

Hence, besides being an alternative for firm growth, internationalisation can be defined as the escalating, continued process of a company getting involved in operations in other countries outside its original base.[16]

Economic theories on internationalisation mainly explain companies' decision making with respect to manufacturing outside their country of origin; in other words, which internal and external factors companies take into account when they consider making direct investments in other countries.[17]

The *market power theory* developed by Hymer in 1960 helps to explain the purpose of control when companies decide to make direct foreign investments (FDI). According to this theory, companies are agents seeking market power. Market imperfections allow companies to branch out internationally, obtaining comparative advantages in relation to local companies in the countries where they operate. Hence, there are two main reasons for a company to control markets abroad. The first is to remove the competition through mergers or collusion. The second relates to using some advantage that is unique to the company, such as privileged access to production factors, better distribution systems, ownership of a unique product, or control of more efficient forms of production so that the company that owns such advantage can claim its economic return.[18]

Taking internationalisation studies further, the *eclectic paradigm* developed by Dunning[19] synthesises and combines especially the economic approaches arising out of the transaction cost studies developed by Williamson in 1975, and demonstrates that the economic cost is not the only factor influencing the decision to invest internationally. This theory explains internationalisation by taking into consideration the characteristics of com-

15 Penrose (1959); Guimarães (1982).
16 Goulart et al. (1994).
17 Hemais (2004); Amatucci (2009).
18 (Ibid.).
19 Dunning (1980).

panies and markets encouraging FDI. According to this theory, companies only make the FDI when they are able to bring together three separate sets of advantages at the same time.[20] The advantages are: a) *Ownership advantages*, that is, advantages related to the organisation's own abilities that it has developed (technologies, knowledge, information, brands, patents, processes, and others) and which allow the company to better position itself in the foreign market (in relation to local or foreign manufacturers). Hence, they include advantages related to tangible and intangible assets; b) *Internalisation advantages*, that is, those advantages that emanate from the company's ability and interest in profiting from the specific advantages it has, instead of assigning them to other companies, which leads the company to operate through a subsidiary and substitute exports for production in the foreign market; and c) *Location advantages*, that is, advantages related to the location where the company plans to install itself and offered by a country or a given area (such as infrastructure, labour costs, and tax incentives and subsidies, among others). Hence, according to this theory, a company may, for instance, have an ownership advantage and an advantage arising out of the internalisation of the activity where it has such specific advantage. At same time, it may not obtain advantages related to the foreign market where it plans to operate. In that case, the company has options other than setting up a subsidiary in the market in question. It is possible that such a company chooses to export or even set up a local distribution centre.

E. Assets ownership

A market's competition scenario may change production settings and lead companies to seek partnerships, mergers, and acquisitions, besides creating other structures like holdings or trading companies or even taking the company publicly. By so doing, companies seek competitive advantages in other countries.

The work of Dunning & Lundan[21] reflects the trend in recent years of a higher FDI being made through mergers and acquisitions. Gammeltoft[22] evaluated foreign direct investment (FDI) in emerging countries, especially the ones making up BRIC (Brazil, Russia, India and China), and high-

20 Dunning (1980 and 1993); Amatucci (2009).
21 Dunning & Lundan (1998).
22 Gammeltoft (2008).

lights that advancing globalisation makes foreign investment increasingly more important. Chen & Hsu[23] examined the effects of internationalisation and resource allocation on company performance, and argue that resource allocation plays an essential role in determining a company's international growth. The authors reported that the companies examined in their study transferred R&D resources to sales, and increased the application of resources in marketing to improve their international performance.

With respect to the overall Brazilian agribusiness scenario, the presence of foreign capital in the country through acquisitions of national companies or shares in business groups has been expanding and the quick concentration has been increasing. Jank et al.,[24] who compared the sugar and ethanol industry to other industries, believe that the first is still not very concentrated owing to the insufficient penetration of foreign companies, and that the existence of disparities implies that there are a number of inefficient plants that are ready for mergers or acquisitions by more efficient companies. That same study found that the companies which have survived the gruelling competition have been making acquisitions and mergers, and that greater concentration decreases costs as a result of the expansion into scale and scope economies, leads to more efficient logistics, and even transfers the primary production to areas where land, labour and taxes are cheaper.

F. Method and procedures

So as to reach the goal set for this article, which is to analyse the concentration and internationalisation process carried out by the ethanol supply chains in the state of São Paulo, responsible for 60% of the Brazilian ethanol production, this study is divided into three stages: (1) calculating the concentration in the industry and identifying the main companies; (2) analysing the internationalisation process followed by the main companies; and (3) identifying and analysing the main aspects related to the changes in ownership of the main companies in the industry

23 Chen & Hsu (2009).
24 Jank et al. (1999).

I. Stage 1: Concentration in the ethanol industry

In the long run, the ambition of companies is to hold the monopoly in the markets where they operate. Such monopoly can be reached through (radical) innovations that put up entry barriers or via industry concentration arising out of mergers and acquisitions.[25] In this study, to identify the existence of concentration in the ethanol market, the concentration rate (CR_K) index was used. With this index, the concentration related to the amount manufactured by each company and was defined as the sum of the market shares held by each one of them. Bain[26] ranked the different concentration levels according to Chart 1. The chart shows CR_4 or CR_8 concentration patterns, ranging from very high concentration all the way to a low one.

Chart 1: Industry concentration patterns[27]

Percentage of the market held by the 4 largest companies	Percentage of the market held by the 8 largest companies	Degree of concentration
75% or higher	90% or higher	Very high
65 to 75%	85 to 90%	High
50 to 65%	70 to 85%	Moderately high
35 to 50%	45 to 70%	Moderately low
35% or lower	45 or lower	Low

With respect to the main ethanol-producing states in Brazil, the state of São Paulo accounts for 50% of the total manufactured in the country, and that state's supply chains are the object of this study.

II. Stage 2: Analysing the industry's internationalisation

First of all, secondary data on the ethanol industry's internationalisation in Brazil were collected. The sources used comprised both Brazilian institutions and companies (MAPA, MDIC, SIAMIG, VALOR ONLINE). Furthermore, four industry specialists, namely Matheus Consoli, Vinícius Trombim, Marcos Fava Neves and Eduardo Leão were interviewed. These

25 Bain (1954).
26 Bain (1968).
27 Source: Bain (1968).

experts were chosen owing to their extensive knowledge of the sugar-alcohol industry and numerous papers published on the subject. Eduardo Leão, specifically, was chosen because he is the person in charge of affairs related to the industry's internationalisation at ÚNICA (União da Indústria de Cana de Açúcar – the sugarcane industry association). Executives from two distilleries producing ethanol were interviewed in order to identify the strategies and processes of internationalisation within the ethanol sector.

III. Stage 3: Changes in ownership

Changes in ownership based on the CR_4 index were evaluated. Once such information on the main companies was obtained, the changes that had taken place during the ownership period were identified. The information found per company was complemented by publications provided by BMF-BOVESPA.[28]

G. Results

Next, the results obtained from the three analyses were presented, namely in respect of concentration, industry analysis and internationalisation, and changes in ownership.

I. Market concentration

The method standards adopted in this study followed a descriptive statistical logic based on the concentration rate (CR_K) offered by Bain.[29] At first, the CR_8 and CR_4 were calculated, both for the 2006/07 harvest; then the CR_8 and CR_4 were calculated for the 2007/08 and 2008/2009 harvests. The purpose of having the concentration rates for those three harvest years was to compare the industry dynamics in the past three years.

Next, in order to obtain more robust results, the Herfindahl-Hirschman (HH) rate was chosen as a second method to find out what the market concentration dynamics had been like in the 2008/09 harvest year in comparison with the 2007/08 and 2006/07 years. The information compiled

28 BM&FBOVESPA (2009).

29 Bain (1968).

focused on the market share held by the eight main economic groups responsible for the largest production in the ethanol market and located in the state of São Paulo in each year. The data were provided by the Bioenergy Producers Association (União dos Produtores de Bioenergia (UDOP))[30] and the Sugarcane Industry Association (União da Indústria de Cana de Açúcar (UNICA).[31]

In Table 1, it is evident – by using the CR_4 index for the 2006/07 harvest – that the four largest groups held 21.83% of the ethanol production. On the other hand, the evaluation of the eight largest companies (CR8) found that the rate was 32.03%. When comparing those data against Bain's chart,[32] it is clear that both rates show that the industry can be ranked at a low level of concentration.

Table1: Ethanol production in the state of São Paulo per economic group, including the 2006/07 harvest concentration rates[33]

Economic group	Production (1,000 litres)	Percentage (%)	Cr4 (%)	Cr8 (%)	H-H rate
Crystalsev	1,160,103	10.63%	10.63%	10.63%	0.01131
Irmãos Biagi	426,320	3.91%	3.91%	3.91%	0.00153
Zilor	402,011	3.68%	3.68%	3.68%	0.00136
São Martinho	393,530	3.61%	3.61%	3.61%	0.00130
Tereos	309,072	2.83%		2.83%	0.00080
Virgolino de Oliveira	305,283	2.80%		2.80%	0.00078
José Pessoa	261,548	2.40%		2.40%	0.00057
Nova América	237,055	2.17%		2.17%	0.00047
Subtotal	**3,494,922**	**32.03%**			
Others	**7,415,091**	**67.97%**			
Total	**10,910,013**	**100.00%**	**21.83%**	**32.03%**	**0.01813**

Table 2 presents the ethanol production data per economic group in the state of São Paulo for the harvest of 2007/2008.

30 UDOP (2009).
31 UNICA (2009).
32 Bain (1968).
33 Sources: UNICA (2009); UDOP (2009).

Table 2: ethanol production in the state of São Paulo per economic group, including the 2007/08 harvest concentration rates[34]

Economic group	Production (1,000 litres)	Percentage (%)	Cr4 (%)	Cr8 (%)	H-H rate (%)
Cosan	1,510,612	11.32%	11.32%	11.32%	0.0128
Santelisa	724,248	5.43%	5.43%	5.43%	0.0029
São Martinho	481,800	3.61%	3.61%	3.61%	0.0013
Zilor	442,860	3.32%	3,32%	3.32%	0.0011
Pedra Agroindustri-al	429,768	3.22%		3.22%	0.0010
Moema	425,974	3.19%		3.19%	0.0010
Tereos	394,371	2.96%		2.96%	0.0009
Moreno	379,797	2.85%		2.85%	0.0008
Subtotal	**4,789,430**	**35.89%**			
Others	**8,555,777**	**64.11%**			
Total	**13,345,207**	**100.00%**	**23.68%**	**35.89%**	**0.0219**

Table 3 presents the ethanol production data per economic group in the state of São Paulo for the harvest of 2008/2009.

Table 3: Ethanol production in the state of São Paulo per economic group, including the 2008/09 harvest concentration rates[35]

Economic group	Production (1,000 litres)	Percentage (%)	Cr4 (%)	Cr8 (%)	H-H rate (%)
Cosan	2,062,701	12.34	12.34	12.34	0.0152
Santelisa	866,699	5.18	5.18	5.18	0.0027
Pedra Industrial	586,794	3.51	3.51	3.51	0.0012
Moreno	568,844	3.40	3.40	3.40	0.0012
São Martinho	568,627	3.40		3.40	0.0012
Zilor	558,626	3.34		3.34	0.0011
Tereos	495,888	2.97		2.97	0.0009
Moema	460,603	2.75		2.75	0.0008
Subtotal	**6,168,782**	**36.89%**			
Others	**10,553,697**	**63.11%**			
Total	**16,722,479**	**100.00%**	**24.43%**	**36.89%**	**0.0396**

34 Sources: UNICA (2009); UDOP (2009).
35 (Ibid.).

By comparing the 2006/07 harvest CR_4 rate against the 2007/08 and the 2008/2009 harvests, the concentration level went from 21.83% to 23.68% and 24.43%, respectively. With respect to the CR_8, it went from 32.03% in the 2006/07 harvest to 35.89% in the 2007/08 harvest and to 36.89% in the 2008/2009 harvest.

Despite both results (CR_4 and CR_8) indicating a low level of concentration, when the three harvests (2006/07, 2007/08 and 2008/2009) are compared a movement towards concentration in this market can be seen. Tables 1, 2 and 3 also show that the production scale is growing owing to the industry's concentration. This may throw up barriers to the entry of new companies if they fail to reach the economic scales currently being set up by a more concentrated industry.

It is important to remember that, in early 2000, the authorities' attention to alternative energy sources was a constant presence. In Brazil, besides the concern regarding the need to reduce greenhouse gas emissions which drove public policies to increase the hydrated ethanol blend to a maximum of 25%, the development of the 2003 model of new flex-fuel engine technology came to favour the national industry of ethanol that consequently sought larger market shares.

The theoretical framework presented earlier states that one of the strategies followed by companies to concentrate a greater market share is mergers and acquisitions (M&A). The changes in the names of the groups and in the production volumes presented in Tables 1, 2, and 3 show that in the São Paulo ethanol market movements pointing toward such a strategy can be seen. To confirm these findings, the H-H rate was used in the study.

The Herfindahl-Hirschman (H-H) rate for the 2006/07 harvest was 0.01813, going up to 0.0219 in the 2007/08 harvest, and to 0.0396 in the 2008/09 harvest (Tables 1, 2, 3), which means a 218% increase from 2006/07 to 2008/09. Although the value is far from 1 (representing a monopoly structure), it shows considerable growth and some evidence that those companies are moving towards market concentration to obtain higher power to establish prices and consequently higher performance. For a more robust conclusion about the movement of concentration, or not, in the ethanol sector, it will be necessary to follow up on the next harvests.

Proença verified the CR4, CR8 and Hirschmann-Herfindahl of sugarcane companies in the state of São Paulo for the harvest of 1999/2000 and 2009/2010, as presented in Table 4 and it corroborates the presented data.[36]

---

36 Proença (2012).

Table 4: Concentration rates – Sugarcane companies in the state of São Paulo[37]

Harvest	Cr4 (%)	Cr8 (%)	H-H
1999/2000	14.6	25.2	181
2005/2006	27.1	37.4	351
2008/2009	25.1	36	314
2009/2010	31.1	40.6	

By comparing the 2008/2009 harvest CR_4 and CR_8 rates against the 2005/06 harvests, the concentration level was lower than those of the other harvests. It might be said that in 2008/2009 new players entered into the market decreasing the concentration rate. However, in 2009/2010 harvest, the concentration rate changed again. It could be said that this was due to new mergers and acquisitions in those years, which increased the productive capacity of the companies, changing the concentration rate again.[38]

In 2010 Brazilian public policy towards ethanol determined a reduction of the ethanol blend percentage to 20%, and the market experienced shortages, which had some impact on the concentration level of the companies and also favoured the entry of international capital into the sector.

Table 5 presents that in 2015/2016 harvest the sugarcane processed from the 25 largest economic groups was 332,471,065 tons, and the eight largest economic groups were responsible for 58.47% of the total sugarcane processed.[39]

Table 5: Economic groups and processing of sugarcane – Harvest 2015/2016[40]

Economic group	Processing (tons of cane)	Percentage (%)
Raízen	57,000,000	17.14
Biosev	28,315,000	85.2
Odebrecht	23,759,000	7.15
São Martinho	20,910,961	6.29

37 Source: Proença (2012).
38 Proença (2012).
39 *JornalCana* (2015).
40 Source: *JornalCana* (2015).

Guarani	20,200,000	6.08
Usacúcar – Santa Terezinha	17,220,856	5.18
Tércio Wanderley	13,571,953	4.08
Lincoln Junqueira	13,418,326	4.04
Total – 8 largest	**194,396,096**	**58.47**
Total – 25 largest	**332,471,065**	**100**

Although there are no data for calculating CR_4, CR_8 and HH rates for the economic groups presented in Table 5, it can be seen that Raízen processes almost double the amount of sugarcane processed by the second largest player, Biosev.[41]

According to Reuters, the sugarcane processed in the 2015/2016 harvest (in the central and southern parts of Brazil) was 617.65 million tons and 28 billion litres of ethanol were produced from this harvest.[42]

II. Internationalisation of the ethanol industry

According to Rothkopf[43], after a few decades of investment by the government and companies in the sugarcane chain, more precisely in ethanol, Brazil has positioned itself as a worldwide leader in the efficient production of this biofuel. Some of the characteristics that have allowed Brazil's worldwide leadership in the biofuel field are the industry's technological progress, such as genetic improvements to sugarcane varieties, sugarcane genome mapping, harvest mechanisation, industrial automation, and the introduction of flex-fuel engines into cars manufactured in Brazil.[44]

Besides the high yield of the sugarcane ethanol made in Brazil (6,800 litres/ha), compared to other important players like the United States (3,100 litres/ha of corn ethanol) and the European Union (3,100 litres/ha of beet ethanol), it can be seen that Brazilian production costs are an advantage, because one litre of Brazilian ethanol costs US$0.17, against US$0.32 in the United States and US$0.56 in the European Union.[45]

41 (Ibid.).
42 Reuters (2016).
43 Rothkopf (2007).
44 Macedo (2007); Marcoccia (2007).
45 Carbini & Marjotta-Maistro (2007); Jank & Rodrigues (2007); Macedo (2007); Marcoccia (2007).

According to MAPA,[46] in 1989, Brazil exported 39,000 m^3, in the amount of US$9.2 million. In 2008, Brazil exported 5,124,000 m^3, equivalent to US$2.39 billion (FOB), divided among the 153 exporting companies, thus becoming the top exporting country in the world. That same year, the United States was the major importer of Brazilian alcohol – importing a volume of 1.53 billion litres.[47]

In 2013, the Brazilian export of ethanol amounted to 1.4 billion litres, of which 0.728 billion litres were imported by the United States in that year, representing 52% of the total exported from Brazil. South Korea, Japan and Nigeria are other important markets for Brazil, representing 0.417; 0.91 and 0.045 billion litres, respectively. As presented in Table 6, the state of São Paulo (SP) accounted for 92.26% of the total ethanol exported in Brazil for the harvest 2015/2016.[48]

Table 6: Ethanol exports from Brazilian states (billion litres) – Harvest 2015/2016[49]

Ranking	State	Volume	Share (%)
1	São Paulo	**1,992**	**92,26%**
2	Minas Gera	0,09	4,17%
3	Paraíba	0,022	1,02%
4	Alagoas	0,021	0,97%
5	Outros	0,034	1,57%
	Total	**2,159**	100,00%

The Brazilian market opened in the 1990s. During the past few years, some industries have been going through a wave of mergers and acquisitions. According to the Minas Gerais Association of Alcohol and Sugar Manufacturers (Sindicato da Indústria de Fabricação do Álcool e do Açúcar de Minas Gerais (SIAMIG)), the share of foreign capital in the sugar-alcohol industry has leapt from below 1% at the beginning of this decade to the current 16%.[50]

According to studies carried out in the industry, the factors helping to attract an intense flow of capital into the industry are: the abundance and

46 MAPA (2009).
47 (Ibid.); MDIC (2009).
48 ANP (2015).
49 Source: UNICA (2009).
50 SIAMIG (2009); Valor Online (2008 and 2010).

quality of land; favourable weather; low farm prices in the new areas chosen to expand the sugar-alcohol industry (compared to international market prices); low labour costs; and foreign companies interested in acquiring the technological expertise developed by Brazil in this field.[51]

The experts interviewed agree with those studies and added that foreign investments also seemed to derive from the following: access to the feedstock boasting the highest conversion into the end product when compared to feedstock from other countries; growing demand in domestic and foreign markets; the companies' access to the foreign market as they already operated in the agricultural commodities business; Brazil's comparative advantages like agriculture-favourable soil and weather; foreign companies interested in obtaining the ethanol production technology held by Brazilian companies.

The highest investments in the industry are made by the companies Louis Dreyfus, Tereos, Abengoa, and publicly-traded companies like Cosan and São Martinho.[52] Additionally, with the exception of the companies Abengoa and Tereos, those seem to be the companies with higher market concentration (considering the acquisition of the company Santaelisa Vale by foreign company Louis Dreyfus, as shown later in this chapter). In December 2009, the Bunge company acquired Grupo Moema, increasing the share of foreign capital in the Brazilian sugarcane industry to approximately 16%.[53] The president of UNICA foresaw that if the negotiations in progress became effective, the share of foreign capital in the ethanol sector would reach 22% in 2010. In 2007, the share of foreign capital in the sector was 7%.[54]

In February 2010, Royal Dutch Shell agreed to enter into a US$12 billion joint venture with the largest Brazilian ethanol producer, Cosan, to become the second largest fuel distribution network in Brazil, just after Petrobras Distribution. The joint venture includes development and production of cellulosic fuel in Brazil.[55]

Hence, as predicted by the *market power theory*, the prospect of those companies reaching an oligopoly situation in the long term seems likely. Removing the competition via mergers and acquisitions has shown to be the way that the aforementioned companies have been operating in the Brazilian ethanol market, which has not cemented itself yet. The integra-

51 Pasin & Neves (2001); VALOR ONLINE (2008).
52 SIAMIG (2009); Valor Online (2008).
53 Valor Online (2010).
54 Jank (2010).
55 Regalado (2010).

tion of foreign and Brazilian companies grants them the following: privileged access to the factors of production; use of those companies' advantages, like the ability to distribute in the international market; and increased production scale, allowing those companies to have greater product volumes to compete both in the Brazilian and foreign ethanol markets.[56]

On the other hand, the set of three advantages leading companies to invest in foreign companies, as pointed out by the *eclectic paradigm*, can also be found in this industry's internationalisation. With respect to ownership advantages, foreign companies' expertise in agricultural commodities can be especially highlighted; with respect to internationalisation, it seems that those companies are interested in integrating the production as a way to profit from the returns in the domestic and foreign markets, once they are both growing markets. Finally, it seems clear that making use of Brazil's comparative advantages for the production of ethanol in relation to other countries is favourable to those companies' competition in the industry.[57]

Hence, both theories seem to explain part of the industry's internationalisation in the country, and there seems to be a trend toward market concentration leading companies to work on a larger production scale, which increases the use of those companies' oligopolic advantages both in the Brazilian and foreign markets.

III. Changes in ownership

The ownership structure of economic groups operating in the sugar and ethanol industry is important in showing whether the industry has been attracting investments from other industries. For instance, foreign investments that change groups' structures via mergers and acquisitions make companies bigger and able to enter and operate in global markets.

According to the concentration rate (CR_4) calculated for the state of São Paulo, Brazil, in the 2007/2008 harvest, the ownership of the four major economic groups is as follows:

56 Hemais (2004); Amatucci (2009).
57 Dunning (1980 and 1993); Amatucci (2009).

1. Cosan-Shell-Raízen

According to sources on its website and BMFBOVESPA information, 60.8% of the Cosan S.A. group's shares are held by the company Cosan Limited and 39.1% by Novo Mercado (Bovespa Brasil). Regarding the shares of Cosan Limited, its 17 December 2009 position showed that 28.61% belongs to Queluz Holdings Limited, a company in the group controlling Cosan, while another part belongs to a different plant in the industry. However, other significant shares belong to capital funds totalling 26.57% of the shares held by shareholders owning over 5% of the shares controlled by investment funds. In other words, the presence of investment funds making up the economic group's ownership is a practice adopted by these companies.

Cosan is considered the world's top sugar exporter and the leading company in the Brazilian market, holding a 9% share in the sugar market and 7% in the alcohol market. The company's shares have also been traded on the New York Stock Exchange (NYSE) since the company went public in August 2007. Since then, the company has secured US$1.2 billion, allowing the Cosan group to implement its growth strategy as it sought to become a global player in the ethanol and sugar industry. Other factors that have led to its changes in ownership are: operating in the world's main consumer and producer markets; reaching the financial capability to make the most of acquisition opportunities; investing in organic growth in various markets and areas; improving the group's positioning as a renewable energy provider; and seeking capital cost reductions.[58] Since 2011, Shell and Cosan signed a joint venture named Raízen.[59]

2. Copersucar

Copersucar was created in 1959 and it has 35 partner mills belonging to 21 economic groups. A source has noted: "In 2008, the cooperative members created Copersucar S.A., a private firm, to gain the flexibility to operate in national and international markets and to grow with new commercial strategies."[60]

58 COSAN (2009).
59 COSAN (2016).
60 Fava Neves et al. (2016).

In 2012, the company expanded its international operation with the opening of two new subsidiaries: Copersucar Asia (Hong Kong) and Copersucar North America (Tennessee, USA), which are added to Copersucar Europe (Rotterdam, The Netherlands). Copersucar North America also owns shares in Eco-Energy, which trades biofuels in the United States. In October 2014, Copersucar and Cargill created Alvean Sugar SL, a 50/50 joint venture specialising in originating, commercialising, and trading raw and white sugar globally.[61]

"Today, it's the world's largest originator and trader of cane sugar, and one of the largest ethanol trading organizations," according to Fava-Neves et al.[62]

3. Santelisa Vale – LDC – Biosev

In October 2009, the controlling interest of Brazil's second largest sugar-alcohol group (Santelisa Vale) was officially transferred into the hands of French multinational Louis Dreyfus Commodities (LDC). In 2016, LDC-SEV was created from the merger of LDC Bioenergia and Santelisa Vale. In 2013, the company changed its name to Biosev, and LDC controls a 65% stake in the merged business.[63]

4. Guarani – Tereos

In 2002, the French sugar conglomerate Tereos acquired the company Guarani, and in 2011 Petrobrás bought a 46% stake.[64]

5. São Martinho

The majority of the group's shares are held by members of the Ometto family and 35.36% of shares are free float and traded in the market via BMFBOVESPA.[65]

---

61 See http://www.cargill.com/news/releases/2014/NA31675459.jsp, accessed 15 November 2016.
62 Fava Neves et al. (2016).
63 Reuters (2016).
64 McKay et al. (2015).
65 São Martinho (2009).

A recent study shows the corporate control of Brazilian sugarcane, as presented in Table 7.

Table 7: The corporate control of Brazilian sugarcane[66]

Company	Share of Brazilian market
Copersucar S.A (2014 – merger with Cargill, 50 per cent owned by each company)	Sugar-Ethanol complex (inc. transportation, storage, processing commercialization): 22 per cent Sugar sales: 19 per cent Ethanol sales: 16.3 per cent
Raizen (2011 merger - Royal Dutch Shell and Brazilian conglomerate Cosan)	Sugar-ethanol complex: 9.5 per cent Sugar: 11 per cent Ethanol: 8.8 per cent
Biosev (2009 merger – Louis Dreyfus Commodities sister company after merging with Brazilian sugarcane producer and processor Santelisa Vale	Sugar-Ethanol complex: 7 per cent
Odebrecht Agroindustrial (Brazilian-based engineering, construction, and chemical conglomerate)	Sugarcane: 6.8 per cent Ethanol: 13.2 per cent
Guarani (Acquired by French sugar conglomerate Tereos in 2002; Petrobrás buys 46 per cent stake in 2011)	Sugarcane: 3.4 per cent Sugar: 5.2 per cent Ethanol: 3.8 per cent
Bunge (US-based food processing corporation)	Sugarcane: 3.6 per cent
Adecoagro (Argentina-based company heavily backed by US billionaire George Soros)	Sugarcane: 2.9 per cent

H. *Concluding remarks and managerial implications*

The purpose of this study was to analyse the ethanol market dynamics, more specifically the aspects related to the industry's concentration and internationalisation dynamics, as well as to give a brief historical summary of the legal and regulatory framework, established through public policy, that might also influence the market dynamics. The quantitative analysis enabled the building of an overall view of the ethanol industry's behaviour in the state of São Paulo and clarified some of its economic characteristics from the standpoint of industrial economics.

The results of the study show that for the CR_4, CR_8, and H-H concentration rates there have been changes in the ethanol industry's manufacturing concentration in the state of São Paulo, pointing to a concentration trend. The CR_8 varied from 32.03% to 36.89% (2006/07–2008/09) The H-H varied from 0.01813 to 0.0396, which means it more than doubled in the same period.

66 Source: Adapted from McKay et al. (2015).

Different pieces of information reviewed recently indicate that the market continues in a dynamic process towards greater concentration, probably owing to its attractiveness, which is generating higher merger and acquisition flows as well as changing competition patterns.

It should be remembered that attractiveness here is meant from the capital perspective, since usually the best period to concentrate, to reduce costs or to enter markets could easily be the *down* periods of the sector, when the assets are cheaper and economy is an imperative.

Qualitatively speaking, considerable changes were found regarding which groups ruled the market in the period analysed, and large companies were even being incorporated by foreign groups. For a more robust conclusion about the movement of concentration, or not, in the ethanol sector it will be necessary to follow up on the next harvests.

In the last few years, the Brazilian government has put in place public policies to combat inflation, which has hit the ethanol sector hard. By capping gasoline costs and limiting the prices of sugar-based ethanol, government has discouraged the sector, influencing the changes of the main market players and the entrance of international capital into the sector.

According to the theoretical definitions provided by economic theories, the ethanol industry is beginning to feature significant oligopolistic characteristics and is evolving towards an increasingly greater concentration, especially around the eight leading business groups. Such a situation is predicted by the theory, as Bain explains.[67] Bain finds that in industries where the economic importance of large manufacturing units (responsible for a substantial market share) is considerable, as is the specific case of the ethanol industry in Brazil, there is a trend toward industrial concentration and entry barriers arise owing to expanding economic production scales.

Ownership data show that the four largest business groups in the São Paulo sugar and alcohol industry have shares traded in the stock market, where they seek to secure investments in their quest to become global players. The percentage of shares held by investment funds in companies like Cosan is substantial. In other groups, like Santelisa Vale, shares have been bought by foreign groups – which shows how attractive the industry is to foreign capital, as it seeks, through the acquisition of these large groups, to control other large plants manufacturing sugar and alcohol in Brazil.

Hence, with respect to the industry's internationalisation, it can be seen that companies need to work on large production scales to operate both in

67 Bain (1968).

the Brazilian market (which is still pulverised and facing a growing demand) as well as more intensely in the foreign market later on. Therefore, waves of mergers, acquisitions and foreign investments in the Brazilian industry have taken place over the past few years as a way to obtain such scale gains, and this study especially highlights the moves made by worldwide companies trading in agricultural commodities, like Louis Dreyfus, Tereos and Bunge. The study also highlights the use of Brazilian comparative advantages (the edaphoclimatic conditions of the country) and the technological mastery developed by the industry as sources of the interest of foreign groups in investing in the country.

The data and interviews also show that the industry faces the great challenge of turning ethanol into an international commodity. That requires the convergence of technical and commercial specifications among importing and producing countries. Today, the specifications defined by each importing country (United States, European Union, Japan, China, India, etc.) prevail, and they are not necessarily equivalent. Standard specifications will allow a great number of producers and buyers to emerge, bringing with them the reliability and efficiency that the major commodity markets have been achieving.

Bibliography

ANP / Agência Nacional do Petróleo, Gás Natural e Biocombustíveis, 2015, *Anuário estatístico brasileiro do petróleo, gás natural e biocombustíveis*, Brasília, Brasil.

Amatucci, M, 2009, "Teorias de negócios internacionais e a economia Brasileira – de 1850 a 2007", in: Amatucci, M (ed), Internacionalização de empresas: teorias, problemas e casos, São Paulo, Atlas, Parte 1.

Anand, BN & T Khanna, 2000, Do firms learn to create value? The case of alliances, *Strategic Management Journal* 21(3), 295.

Bain, JS, 1951, Relation of profit rate to industry concentration: American manufacturing, 1936 – 1940, *The Quarterly Journal of Economics* 65(3), 293.

Bain, JS, 1954, economies of scale, concentration, and the condition of entry in twenty manufacturing industries, *The American Economic Review* 44(1), 15.

Bain, JS, 1968, *Industrial organization*, University of California, Wiley Edict, Berkeley.

Barney, J, 1991, Firm resource and sustained competitive advantage, *Journal of Management* 17, 97.

BM&FBOVESPA, 2009, *Empresas listadas*, at http://www.bmfbovespa.com.br/cias-listadas/empresas-listadas/BuscaEmpresaListada.aspx?idioma=pt-br, accessed 17 December 2009.

Cabrini, MF & MC Marjotta-Maitro, 2007, Mercado internacional de álcool: os recentes programas de uso do produto como combustível, *Agroanalysis,* 7 March, 36.

Chen, H & CW Hsu, 2009, Internationalization, resource allocation and firm performance, *Industrial Marketing Management* 39(7), 1103.

Coase, R, 1937, The nature of the firm, *Economica* 16(4), 386.

COSAN, 2009, *Annual Report 2008*, at http://www.cosan.com.br/cosan2009/web/arquivos/Cosan_RelatorioAnual_20090227_port.pdf, accessed 17 December 2009.

COSAN, 2016, Destaque em energia, at http://cosan.com.br/pt-br/negocios/energia/raizen, accessed 15 November 2016.

Dunning, JH, 1980, Toward an eclectic theory of international production: Some empirical tests, *Journal of International Business Studies* 11(1), 9.

Dunning, JH, 1993, "Global strategy and the theory of international production", in: Dunning, JH, *The globalization of business*, London, Routledge.

Dunning, JH & SM Lundan, 1998, The geographical sources of competitiveness of multinational enterprises: An econometric analysis, *International Business Review* 7(2), 115.

Dyer, J & H Singh, 1998, The relational view: cooperative strategy and sources of interorganizational competitive advantage, *Academy of Management Review* 38(4), 660.

Fava Neves, M, AW Grayb & BA Bourquardi, 2016, Copersucar: A World Leader in Sugar and Ethanol, *International Food and Agribusiness Management Review* 19(2), 207, at https://www.ifama.org/resources/Documents/v19i2/1020150044.pdf, accessed 15 November 2016.

Ferguson, PR & GJ Ferguson, 1994, *Industrial economics: issues and perspectives*, New York, NYU Press.

Gammeltoft, P, 2008, Emerging multinationals: outward FDI from the BRICS countries, *International Journal of Technology and Globalisation* 4(1), 5.

Goulart, L, CA Arruda & HV Brasil, 1994, A evolução na dinâmica de internacionalização, *Revista Brasileira de Comércio Exterior* 41, 31.

Grant, RM, 1991, The resource-based theory of competitive advantage: implications of strategy formulation, *California Management Review* 33(3), 114.

Guimarães, EA, 1982, *Acumulação e crescimento da firma: um estudo de organização industrial*, Rio de Janeiro, Zahar.

Gulati, R, N Nohria & Zaheer A, 2000, Strategic Networks, *Strategic Management Journal* 21(3), 203.

Hemais, CA (ed), 2004, *O desafio dos mercados externos: teoria e prática na internacionalização da firma*, Rio de Janeiro, MAUAD.

Hou, K & D Robinson, 2006, Industry concentration and average stock returns, *Journal of Finance* 61(4), 1927.

Holmström, B & J Roberts, 1998, The Boundaries of the Firm Revisited, *Journal of Economic Perspectives* 12(4), 73.

Jank, MS, MFP Leme, AM Nassar & PF Filho, 1999, Concentration and internationalization of Brazilian agribusiness exporters, *International Food and Agribusiness Management Review* 2(3), 359.

Jank, MS & L Rodrigues, 2007, Dinâmica e agenda do setor sucroalcooleiro na próxima década, *Revista de Política Agrícola* 16(4), 86.

Jank, MS, 2010, A Globalização e o setor sucroenergégico brasileiro, at HTTP://www.unica.com.br, accessed 26 March 2010.

JornalCana, 2015, Produção, at https://www.jornalcana.com.br/edicao/jornalcana-260/, accessed 6 June 2016.

Macedo, IC, 2007, *Situação atual e perspectivas do etanol, Estudos Avançados São Paulo* 21(59), 157.

MAPA / Ministério da Agricultura, Pecuária e Abastecimento, 2009, *2008 Comércio Exterior Brasileiro*, at http://www.agricultura.gov.br/pls/portal/docs/PAGE/MAPA /ESTATISTICAS/COMERCIO_EXTERIOR_BRASILEIRO/EXPORTA%C7%D5ES% 20BRASILEIRAS%20DE%20ALCOOL%20DEZ-08_0.PDF, accessed 10 April 2009.

Marcoccia, R, 2007, *A participação do etanol brasileiro em uma nova perspectiva na matriz energética mundial*, Dissertação (Mestrado) - Programa Interunidades de Pós-Graduação em Energia, Escola Politécnica, Faculdade de Economia, Administração e Contabilidade, Instituto de Eletrotécnica e Energia, Instituto de Física, São Paulo, Universidade de São Paulo.

McKay, B, S Sauer, B Richardson & R Herre, 2015, The political economy of sugarcane flexing: Initial insights from Brazil, southern Africa and Cambodia, *Journal of Peasant Studies* 43(1), 195.

MDIC / Ministério do Desenvolvimento, Indústria e Comércio, 2009, at http://www.mdic.gov.br/sitio/interna/interna.php?area=2&menu=999, accessed 28 October 2009.

Pasin, RM & MF Neves, 2001, "Mergers, acquisitions and the internationalization of the sugar cane industry in Brazil", in: Third International Conference on Agri-Food Chain/Networks Economics and Management, Ribeirão Preto.

Penrose, E, 1959, *The theory of the growth of the firm*, Oxford, Oxford University Press.

Peteraf, MA, 1993, The cornerstones of competitive advantage: A resource-based view, *Strategic Management Journal* 14(3), 179.

Proença, ER, 2012, Concentração, integração horizontal e vertical das usinas canavieiras, doctoral thesis (doutorado em agronomia), Ilha Solteira, UNESP.

Qualls, D, 1972, Concentration, barriers to entry, and long run economic profit margins, *The Journal of Industrial Economics* 20(2), 146.

Regalado, A, 2010, Race for cellulosic fuels spurs Brazilian research program, *Science* 327(19), 928.

Reuters, 2016, Produção de etanol do centro-sul bate recorde pela 3ª safra seguida, at http://br.reuters.com/article/topNews/idBRKCN0X91N4, accessed 6 June 2016,

Rothkopf, GA, 2007, Blueprint for green energy in the Americas, at http://www.iadb.org/biofuels/, accessed 26 December 2007.

Rumelt, R, D Schendel & D Teece, 1991, Strategic management and economics, *Strategic Management Journal* 12, 5.

São Martinho, 2009, Governança corporativa: Relação com os investidores, at http://www.usinasaomartinho.ind.br/, accessed 17 December 2009.

SIAMIG, 2009, Sindicato da indústria do açúcar no estado de minas gerais. capital estrangeiro no setor sucroalcooleiro brasileiro, 007 *Relatório Econômico* 28 January 2009.

UDOP / União dos Produtores de Bioenergia, 2009, União dos produtores de bioenergia, at http://www.udop.com.br/index.php?item=safras, accessed 28 October 2009.

UNICA / União da Indústria de Cana-de-Açúcar, 2009, União das indústrias de cana de açúcar, available at http://www.unica.com.br/dadosCotacao/estatistica/, accessed 28 October 2009.

Valor online, 2008, Especial – Biocombustíveis, a força do verde, at http://www.valor online.com.br/, accessed 10 December 2008.

Valor online, 2010, Bunge acerta compra de usinas do grupo Moema, at http://www.valoronline.com.br/?online/agronegocios/78/6020268/bunge-compra-usinas-do-grupo-moema&scrollX=0&scrollY=216&tamFonte=, accessed 6 January 2010.

Wernerfelt, B, 1984, A resource-based view of the firm, *Strategic Management Journal* 5, 171.

Williamson, O, 1975, *Markets and hierarchies: analysis and antitrust implications*, New York, The Free Press.

Williamson, O, 1985, *The economic institutions of capitalism*, New York, Free Press.

Williamson, O, 2002, The theory of the firm as governance structure: from choice to contract, *The Journal of Economic Perspectives* 16(3), 171.

Williamson, O, 2010, Transaction cost economics: the natural progression, *American Economic Review* 100(3), 673.

The current status of biofuels to promote energy security – legal challenges and potential for sustainable development in South Africa*

Lena C Plato, LLM candidate Stellenbosch University, South Africa; Criminal Defense Lawyer, Benecken & Partner, Germany

A. Introduction

Various scientific studies have revealed that climate change has started affecting the atmosphere adversely and, of all the continents, Africa in particular is suffering detrimental consequences.[1] South Africa is experiencing an often subtle, yet steady, change in climate, and temperatures have risen significantly over the last 60 years and are predicted to continue their rising trend.[2] Governments all over the world have created policies and measures to reduce the carbon dioxide (CO_2) emissions that are mainly responsible for climate change.[3] As yet, developing African countries have been more concerned with the issues relating to accessing energy in order to improve industrial production and increase economic growth than with crafting policies that reduce carbon dioxide emissions or halt climate change.[4] Compared to other African countries, the South African economy is energy-intensive and the energy consumption rate is the highest in Africa.[5] However, only a small number of primary energy sources exist in South Africa, with limited substitution between energy carriers and an overwhelming reliance on coal.[6] Until the present time, energy efficiency and renewable energy were not considered to be an issue in South Africa.[7]

---

* This chapter is based on the author's LLM thesis conducted under the supervision of Prof Dr Oliver C Ruppel at the Faculty of Law, University of Stellenbosch, South Africa.

1 Lotz-Sisitka & Urquhart (2014: 8); Meyer & Odeku (2009: 49).

2 Griffin (2012).

3 Meyer & Odeku (2009); Johnson et al. (2007).

4 Davidson & Winkler (2003); van Vuuren (2008: 54-55); Danish (2007).

5 Emissions statistics available at http://mdgs.un.org/unsd/mdg/SeriesDetail.aspx?srid=749, last accessed 15 January 2015; Meyer & Odeku (2009: 49).

6 Tyler (2011).

7 Meyer & Odeku (2009: 49).

In 2003 the South African government released a white paper on renewable energy which sought to set a target of producing 10,000 GWh of energy from renewable energy sources, specifically biomass, wind, solar and small-scale hydro, by 2013.[8] Based on the white paper[9] and its acknowledgement of the importance of a diversified energy supply mix, South Africa has also established several biofuel policies and mandates, such as the Biofuels Industrial Strategy of the Republic of South Africa, which was published by the Department of Minerals in December 2007.[10] The aim of this biofuels strategy was to achieve a 2% penetration of biofuels in the national liquid fuel supply, which represents about 30% of the national renewable energy target for 2013, as set out in the white paper on renewable energy.[11]

Biofuels have the potential to provide numerous environmental, and energy security and efficiency benefits for the South African economy, but large-scale biofuel production has not materialised yet.[12] Although biofuels are not new to South Africa, the national biofuels industry is still in its infancy.[13] Entrance restrictions are considerable and the energy market is presently monopolised by a single provider of electricity.[14] The key to success for biofuels development in South Africa is the implementation of a sound legal, policy and regulatory framework that will facilitate large-scale investments in this resource.[15]

---

8 Department of Minerals and Energy (1998).

9 While the Renewable Energy White Paper indicated that biofuels made from biomass were among the renewable energy sources under consideration in order to achieve industry viability, it was clear that the biofuels policy needed to be developed to a much deeper level than was laid out in the white paper.

10 Letete & Blottnitz (2012).

11 Biofuels Industrial Strategy of the Republic of South Africa, Department of Minerals and Energy (2007).

12 There is still no workable biofuel project up and running, despite many policy statements and plans for projects. South African Airways (SAA) will be powering some of its aircraft from fuel derived from tobacco in 2016. See Gilder & Mamkeli (2014).

13 During the past thirty years, two types of renewable fuels, biodiesel and ethanol, have been considered as products that could be produced locally. Biodiesel was considered, particularly by the agricultural industry, as a replacement for diesel in agriculture. The technical developments of manufacturing this fuel and proving its efficiency were conducted in the period 1979 to 1983. See: http://proteinresearch.net/index.php?dirname=html_docs_ 025biofuels, accessed 15 January 2015.

14 Chambers (2006).

15 Odeku (2012).

But development of the biofuels industry has proved difficult and it seems that the national biofuels strategy hampers the establishment and expansion of the industry.[16] This situation is due to a lack of commitment from the South African government and an utter distortion of information on the current policy.[17] This makes a successful, comprehensive legal, regulatory and constitutional framework for the development of biofuels in South Africa all the more important.

B. *South Africa's energy policy*

The South African economy is energy-intensive, meaning that the country uses a large amount of energy for every rand of economic output.[18] Economic growth and the demand for electricity in rural areas have increased the demand for electricity in South Africa.[19] Indeed, the publication released by Statistics South Africa (Stats SA) in 2014, containing the latest data on electricity produced and available for distribution, shows the amount of electricity used in 2013 as 256,073 GWh and an amount of 252,578 GWh in 2014.[20] Despite this small decrease South Africa's energy requirement, the demand is expected to double 2013 usage levels by the year 2030.[21]

The South African energy sector is very simple in structure because there are only a small number of primary energy sources, with limited substitution between energy carriers, and an overwhelming reliance on coal.[22] Coal-fired power generation provides over 80% of the country's electricity supply.[23] The country has large reserves of uranium, but only small reserves of oil and gas – which leads to the need for imports mainly from Saudi Arabia and Nigeria.[24] But dependence on outside sources

16 Funke (2011).
17 (Ibid.).
18 Hughes et al. (2002).
19 Cigrasp (2013).
20 Statistics South Africa available at http://www.statssa.gov.za/?p=4045, accessed 15 January 2015.
21 Cigrasp (2013).
22 Tyler (2011).
23 (Ibid.).
24 36% (Saudi Arabia), 34% (Iran), and 16% (Nigeria) in 2006. Imports from Iran had ceased when Western nations imposed sanctions on that country. In 2013 45% of the oil came from Saudi Arabia and 23% from Nigeria; cf. Nkomo (2009).

leaves South Africa vulnerable to factors beyond its control, such as oil price shocks and supply disruptions due to political and other external factors. Price fluctuations result in a flow of foreign exchange resources to oil producers, which might otherwise have been used to stimulate the local economy. Other energy sources like nuclear and especially renewable energy are produced in such small quantities as to be of no consequence. This being so, South Africa continues its dependence on coal for the production of electricity.[25] This means that the country's economy is highly dependent on fossil fuels, and, as seen above, it is one of the highest emitters of greenhouse gas per capita in the world.[26] Figure 1 illustrates the total energy consumption in South Africa in 2013.

Figure 1: Energy consumption in South Africa in 2013[27]

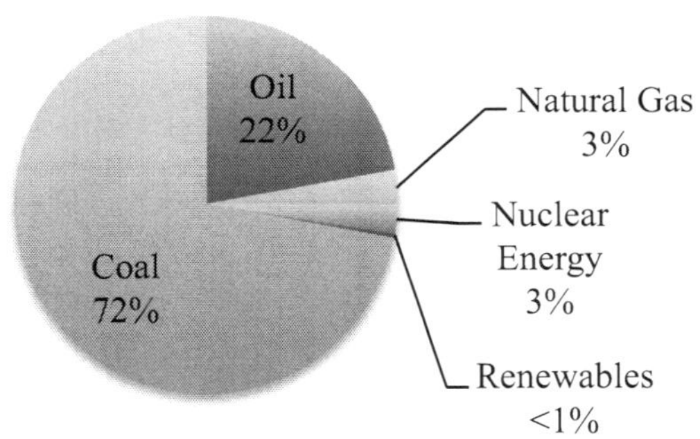

C. *Renewable energy policy in South Africa*

The global energy context shows that fossil fuels still dominate the system in South Africa. Not only in South Africa, but also worldwide, fossil fuels contributed 77.9% to the global energy system.[28] Despite this, renewable energies are steadily becoming a greater part of the global energy mix

25 Tyler (2011).
26 EREC (2011).
27 Source: BP Statistical Review of World Energy (2014).
28 WEC (2013).

with double-digit growth rates being observed in the last decade for some renewable energy technologies.[29]

Most of South Africa's renewable energy is provided by biomass, which consists primarily of fuelwood used in households, which is harvested unsustainably, or of biomass, in the form of sugar cane.[30] Moreover, pulp mills generate electricity directly from bark and 'black liquor'. The wind energy contribution to South Africa's electricity supply is provided by the privately owned Darling Wind Farm in the Western Cape and Eskom's Klipheuwel Wind Farm.[31] With regard to hydropower, South Africa has only 668 MW of installed capacity.[32] Small hydro plants account for about 68 MW of this capacity. In addition, pumped storage schemes with a capacity of 1,580 MW have been installed.[33] South Africa's electricity access has grown by over 50% since 1994, but renewable energy is lagging behind.[34] South Africa's clean energy penetration stands at around 511 MW of installed capacity out of a national total of almost 43 GW.[35] However, in 2013, nearly 180 MW was added, mainly in the form of solar energy, and further expansion is planned.[36]

South Africa's history of using biomass as an energy source dates back to the 1920s when ethanol derived from sugar cane was mixed with petrol.[37] Between the 1970s and early 1990s South African involvement in the development of alternative fuel sources was largely in response to sanctions placed on the apartheid government.[38] There are currently no large-scale biofuel producers in South Africa, but there is much interest in this field in both the agricultural and corporate sectors.[39] Sugar producers and maize farmers represent the majority of the parties looking at driving the South African bioethanol industry.[40] Biodiesel, mostly from recycled sunflower oil, is produced on small scale – less than 300,000 litres per year – by many farmers and small companies across the country; and virgin soybean, and sunflower and canola oils have also been used for bio-

29 (Ibid.).
30 Winkler (2006).
31 Peters (2013).
32 Banks & Schaeffler (2006).
33 (Ibid.).
34 Fripp (2014).
35 (Ibid.).
36 (Ibid.).
37 von Maltitz et al. (2009).
38 Lynd et al. (2013).
39 Department of Minerals and Energy (2007).
40 (Ibid.).

diesel production.[41] In 2014 South Africa consumed about 13.1 billion litres of diesel, and consumption is forecast to grow at 4.5% per year between 2012 and 2020.[42] Biofuels could be a partial solution for the problems mentioned in section 1.2 and could decrease South Africa's dependence on fossil fuels and imported oil, while also decreasing pollution and promoting the renewable energy production ratio. But, in order to promote the introduction and growth of renewable energy technologies, it is important to have government support.[43] The growth rate of such technology is responsive to the energy policy guidelines of a country.[44]

D. Biofuels regulatory framework in the context of climate change and energy security in South Africa

I. Status of the biofuels industry

The increasing demand for energy has created a need for alternative energy sources.[45] In particular, this can be coupled to the large growth in renewable energy generation.[46] Biofuels have been promoted as an environmentally sustainable solution to the global and especially the South African energy crisis.[47] Although biofuels are mainly used to replace or supplement the conventional petroleum-based transportation fuels, they can also be deployed to generate heat and electricity.[48] Being an alternative to fossil fuels, biofuels can be applied to existing vehicles with little or no engine modification and, although they release carbon dioxide when burned in internal combustion engines, they differ from fossil fuels partly because their use reduces the net emission of carbon dioxide and other gases associated with global climate change and they are biodegradable.[49] Developing biofuels has many further advantages such as utilising renewable resources efficiently; enhancing energy security and energy supply diversification; enhancing rural agricultural development and investment

41 Letete (2009).
42 SAPIA (2013: 6).
43 Girard & Fallot (2006).
44 Sebitoso & Pillay (2008).
45 Amigun et al. (2011).
46 Sebisoto & Pillay (2008).
47 von Maltitz & Brent (2008).
48 ResearchandMarkets (2015).
49 (Ibid.).

in rural areas; reducing greenhouse gas emissions; and increasing jobs and improving livelihoods.

Global production of biofuels has been growing steadily from about 20 billion litres (125 million barrels) in 2001 to over 110 billion litres (692.5 million barrels) in 2011.[50] During this period, worldwide production of ethanol and biodiesel has increased almost five- and twenty-fold, respectively. Global biofuel production is projected to reach 222 billion litres by 2021, with an ethanol and biodiesel share of 81% and 19%, respectively.[51] In 2011, biofuels provided around 3% of total fuel for road transportation worldwide and it is projected to share 27% of world transport fuel by 2050.[52] The United States is the top producer of biofuels, followed by Brazil, whereas the global contribution of South Africa to biofuel production is well below 0.01%.[53] Indeed, South Africa's biofuel development has been stalled by a legislative process and biofuel is yet to be commercially produced on a large scale.[54]

Despite many policy statements and plans over the years, biofuel production in South Africa is still in its infancy.[55] Only a very few small-scale biofuel plants are available in the country.[56] South Africa has a much larger capacity for producing bioethanol than biodiesel. Currently, there are about 200 small entrepreneurs who produce biodiesel on a small scale, mostly using waste vegetable oil (WVO) as feedstock, which neither competes with food nor for agricultural land.[57] A major concern of these entrepreneurs is that the feedstock is too expensive. Furthermore, there is no uptake of biofuels because a mandatory blending does not exist.[58] The production rate of these plants is quite low because owners prefer batch reactors over continuous reactors because of low acquisition cost, simple design and ease of operation.[59] Ethanol, which is produced on small and medium scale, is mostly being used for non-fuel purposes.[60] Several large sugar companies like Illovo and Tongaat Hulett are producing bioethanol for the spirits and chemical manufacturing markets, but not for the liquid

---

50 US Energy Information Administration (2013).
51 Pradhan & Mbohwa (2014).
52 (Ibid.).
53 (Ibid.).
54 (Ibid.).
55 Esterhuizen (2009).
56 Avinash et al. (2014).
57 van Zyl & Prior (2009).
58 (Ibid.).
59 Pradhan & Mbohwa (2014).
60 von Maltitz & Brent (2008).

fuels market.[61] The lone producer of fuel bioethanol is Silversands Ethanol, located in Hoopstad in rural North West province, and manufactures bioethanol gel.[62] The gel is thickened bioethanol that can be used as a cooking and lighting substitute for paraffin.[63] Silversands grows and uses grain sorghum as the feedstock for bioethanol production.[64] The 2011 estimate of the production of ethanol and biodiesel in South Africa was only about 16,000 and 4,770 litres per day, respectively.[65]

II. Feasibility and potential of biofuels

Southern Africa – and especially South Africa – has been identified as an area with extensive biofuel potential, when compared to the rest of the globe.[66] Not only does South Africa have a climate suited to high levels of biomass production, but there is also the perception that much of its land is available for biofuel production.[67] The areas which constituted the former homelands in South Africa are suggested as appropriate for biofuel expansion, based on the view that the land is underutilised.[68] Furthermore, commercial farmers are keen on the establishment of a biofuels industry as they see it as a mechanism to boost agricultural production.[69] South African maize farmers, for instance, argue that they can produce 14 million tonnes of maize, but that the local demand is only 9 million tonnes.[70] South Africa has the technical potential to produce maize for biofuels and farmers believe that a bioethanol industry will help to stabilise the local market.[71] A study further revealed that South Africa produces about 18 million tonnes of agricultural and forestry residues, to which can be added

61 Strydom (2009).
62 Silversands Ethanol, see: http://somnium.co.za/Silversands%20Ethanol/index.html.
63 (Ibid.).
64 Grain sorghum was used primarily to avoid food scarcity concerns, and, during its recent expansion, Silversands kept the food security issue at the forefront.
65 Pradhan & Mbohwa (2014).
66 von Maltitz & Brent (2008).
67 (Ibid.).
68 The areas that constituted the former Bantu homelands in South Africa are suggested for biofuel expansion based on the view that the land is underutilised.[For clarity, insert 'Bantu' if you write 'homelands' in lower case (since 'homelands' comes from the Bantu Homelands Act of 1970).
69 von Maltitz & Brent (2008).
70 (Ibid.).
71 (Ibid.).

more than 8 million tonnes of invasive species that would be available on an annual basis for more than a decade owing to abundant seed banks in infested regions.[72] Furthermore, it has been projected that an additional 67 million tonnes of energy crops per annum can be cultivated on only 10% of available land and that the total estimated biomass production capacity of South Africa is 94 million tonnes per annum.[73]

III. The South African Industrial Biofuels Strategy 2007

The South African government developed the Biofuels Industrial Strategy (BIS) to facilitate the development of an alternative energy source and to assist countries to move from a reliance on expensive, imported, carbon-intensive fuels to more diverse energy sources which include biofuels.[74] The BIS, developed in December 2007, signalled the government's support for the development of biofuels and the beginning of the biofuel industry in South Africa.[75] The BIS was developed by members of the Biofuels Task Team, which consisted of ministers from the agricultural, environmental, land affairs, energy, science and technology, and treasury departments and was formulated as a developmental strategy aimed at using biofuels development to achieve goals set forth by the various departments.[76] The aims of the strategy included attracting investment into rural areas, promoting agricultural development, alleviating poverty through sustainable income earning opportunities, and substituting imports of foreign oil.[77] Furthermore, the strategy aimed to achieve 2% penetration of biofuels in the national liquid fuel supply by 2013, which is equivalent to 400 million litres per annum.[78] The 2% target should be achieved through a B2 or 2% blending level for biodiesel and an E8 or 8% blending level for bioethanol.[79] The strategy also estimated the creation of about 25,000 jobs in rural farming.[80] Finally it determined a contribution of up to 50% to the renewable energy target of 10,000 GWh by 2013.[81]

---

72 van Zyl & Prior (2009).
73 (Ibid.).
74 Department of Minerals and Energy (2007).
75 Strydom (2009).
76 (Ibid.).
77 Pradhan & Mbohwa (2014); Department of Minerals and Energy (2007).
78 Brent et al., (2010).
79 Letete & von Blottnitz (2012).
80 Brent et al., (2010).
81 (Ibid.).

IV. Biofuels regulatory framework

The main policy obstacle for biofuel development in South Africa is the absence of an implemented policy that takes into account the problems discussed in section 2 related to energy security and climate change, but also issues related to agriculture and land, water, and food security. There is a pressing need for the promulgation of a biofuels regulatory framework, policy and regulations – which are currently works in progress – to impose standard criteria for investments, targets for blending, and a price policy.

On 15 January 2014 the Department of Energy took the step of publishing its Draft Position Paper on the South African Biofuels Regulatory Framework, which included regulations on the mandatory blending of biofuel with petrol and diesel and a pricing framework. According to the document, the South African government had identified the biofuels industry as a potential major source of employment and economic development. However, the Department of Energy is still in the process of revising the position paper. Thus, the implementation of the draft position paper of January 2014 alone leaves gaps regarding the critical concerns mentioned above that arise in the Biofuels Strategy, 2007. The enormous development in the field of biofuels means that constant observation of the progress in biofuels generation is necessary. This requires role players, including policy makers, to be at the cutting edge of technology and feedstock breeding.

It is important to point out that South African government has already created the foundation for the biofuels industry. However, it fails to clear the last hurdle of resolving the remaining weaknesses in the draft position paper, and finally implementing it. Examples from other countries show what an important role a detailed biofuels regulatory framework can play for a nation's biofuels industry. Brazil is the second largest producer and consumer of bioethanol, and its soils and climate have similarities to those in South Africa. Both Africa and South America are widely recognised as the continents with the greatest potential to increase modern bioenergy production. As in South Africa, energy security is one of the key drivers of the expansion of ethanol and biodiesel production in Brazil, where domestic feedstock production is favoured as a means to substitute fossil fuel derivatives.

V. Brazil's biofuels regulatory framework – a lesson for South Africa

Brazil's success in the production of alcohol fuel is strongly related to the action of the state as a designer of public policies, as well as to the fact that the government has directly influenced the strategic decisions of the sector.[82] The setting up of a regulatory framework, the concession of credit, the establishment of research institutions and the incentives for the emergence of a domestic and international demand for alcohol fuel are some of the main factors that indicate that the government has been a driving force in the establishment of Brazil's biofuel model.[83]

In Brazil the central pillar of the biodiesel regulatory framework is the National Biodiesel Production Programme (PNPB), created by an interministerial working group in July 2003.[84] The framework establishes guidelines pertaining to biodiesel production; certification and marketing for reasons of job increase; reduction of GHG emissions; guaranteeing of energy supply; creation of a competitive market; assurance of quality and supply from different raw material;, increase in exports; the gaining of more space in the international market; and internal and regional development based on feedstock production.[85] In 2004, the Brazilian National Congress drafted a project of law to include biodiesel as a renewable source in the Brazilian energy matrix, which was enacted by Provisional Measure n. 214/2004, and subsequently converted to Federal Law n. 11.097/05.[86] Many of the parameters above were later defined in Federal Law n. 11.116 of May 2005.[87] Moreover, Law n. 11.116 reduced tax percentages for biodiesel import and production, and delegated power to the Executive to further alter the sales and social security tax rates.[88]

Although mainly driven by economic incentives and market-based instruments, the new trend in Brazilian biofuels policies is to promote environmental best practices in its biofuels industry.[89] By incorporating this aspect of sustainability in its policy, Brazil has driven biofuel production

82 Oliveira & Hira (2014).
83 (Ibid.).
84 Teles da Silva & Dutra (2014).
85 (Ibid.).
86 (Ibid.).
87 Federal Law 11.116, available at: http://www.planalto.gov.br/ccivil_03/_ato2004-2006/2005/Lei/L11116.htm, last accessed 15 January 2015.
88 Teles da Silva & Dutra (2014).
89 Brits (2011).

from an idea to a corporeal industry that is supported and protected by public and private sectors in that country.[90]

South Africa already does have an established energy and fuel industry infrastructure with regulatory measures that relate to renewable energy and biofuels specifically, as the white papers and the Biofuels Industrial Strategy have shown.[91] What South Africa should however learn from Brazil is that the constant implementation of legislative and legally binding rules regarding biofuels can have enormous positive implications for the biofuels industry and finally for energy security and climate mitigation.

The recent work of South Africa's Department of Energy regarding biofuels is already enough to fire the starting pistol for a successful biofuels market entrance. It will take a while to see in which direction biofuels production will develop. It is therefore important to monitor the permanent developments in a fluctuating industry like the biofuels industry. Problems which arise can easily be solved through observing the industry and adjusting rules, where necessary. But without any legally binding framework the biofuels industry will come to a standstill, which would be an enormous waste of the potential.

The example of Brazil shows the advantages that a biofuels industry can bring to a country. South Africa can be the role model for other countries on the continent and become the African leader in the biofuels market. The effects of climate change are already being felt and the energy crisis has created sporadic havoc since 2008 through load shedding, resulting in an enormous loss of money for the country. The time for change is now.

E. Conclusion

The current biofuels policy indicates that biofuel production in South Africa cannot and will not take place without an effective policy. The current lack of clarity about the implementation of the biofuels regulatory framework – coupled with a severe lack of commitment from the government, especially from the Department of Energy with respect to blending targets and price policy – is the industry's main drawback. The overarching conclusion of this article is that the development of the biofuels industry in

90 (Ibid.).
91 (Ibid.).

South Africa could make a positive contribution to the development of South Africa's energy sector and to climate mitigation, and will become more important in the future, especially in respect of alternatives to increased fossil fuel combustion. There is a huge potential for biofuels in South Africa, but without further steps by the government, the biofuels industry cannot develop.

Bibliography

Amigun, B, JK Musango & W Stafford, 2011, Biofuels and sustainability in Africa, *Renewable and Sustainable Energy Reviews* 14, 690.

Avinash, A, D Subramaniam & A Murugesan, 2014, Biodiesel – a global scenario, *Renewable and Sustainable Energy Reviews* 29, 517.

Banks, D & J Schaeffler, 2006, *The potential contribution of renewable energy in South Africa*, at http://www.nano.co.za/PotentialContributionOfRenewableEnergyInSA Feb06.pdf, accessed 15 January 2015.

BP / British Petroleum, 2014, *BP statistical review of world energy – June 2014*, London, BP, at http://www.bp.com/content/dam/bp/pdf/Energy-economics/statistical-review-2014/BP-statistical-review-of-world-energy-2014-full-report.pdf; accessed 11 February 2015.

Brent A, R Sigamoney, H von Blottnitz & S Hietkamp, 2010, Life cycle inventories to assess value chains in the South African biofuels industry, *Journal of Energy in Southern Africa* 21(4), 15.

Brits, D, 2011, *The regulatory regime for biofuels in South Africa*, at: http://dspace.nwu.ac.za/bitstream/handle/10394/8098/Brits_D.pdf?sequence=2, accessed 15 January 2015.

Chambers, D, 2006, *Biofuels in South Africa – factors influencing production and consumption*, at http://repository.up.ac.za/bitstream/handle/2263/23702/dissertation.pdf?seq uence=1, accessed 15 January 2015.

Cigrasp, 2013, *Energy profile South Africa*, at http://cigrasp.pik-potsdam.de/countries /833900607/energy_profile, accessed 15 January 2015.

Danish, KW, 2007, An overview of the international regime addressing climate change, *Sustainable Development, Law & Policy* 7(2), 10.

Davidson, O & H Winkler, 2003, *South Africa's energy future: Visions, driving factors and sustainable development indicators*, at http://www.developmentfirst.org/Studies/South AFricaCountryStudies.pdf, accessed 15 January 2015.

Department of Minerals and Energy, 1998, *White paper on the energy policy of the Republic of South Africa*, at http://www.energy.gov.za/files/renewables_frame.html, accessed 15 January 2015.

Department of Minerals and Energy, 2007, *Biofuels industrial strategy of the Republic of South Africa*, at http://www.energy.gov.za/files/esources/renewables/biofuels_indus _strat.pdf(2).pdf, accessed 15 January 2015

EREC, 2011, *The Advanced Energy Resolution: A Sustainable Energy Outlook for South Africa Report* (2nd edition), at http://www.energyblueprint.infofileadminmedia/

documents/national/2011/E_R__South_Africa_May_2011-LR.pdf, accessed 15 January 2015.

Esterhuizen, D, 2013, *Sugar annual – Republic of South Africa*, Global Agricultural Information Network, United States Department of Agriculture, at http://gain.fas.usda.gov/RecentGAINPublications/SugarAnnual_Pretoria_South Africa - Republic of_4-17-2013.pdf, accessed 15 January 2015.

Fripp C, 2014, Renewable energy in South Africa still lacking, but on the right path, 2 December 2014, at http://www.htxt.co.za/2014/12/02/renewable-energy-in-sa-still-lacking-but-on-the-right-path/, accessed 15 January 2015.

Funke, TB, 2011, *Biofuel production in South Africa: the games, the cost of production and policy options*, doctora thesis at http://repository.up.ac.za/bitstream/handle/2263/24639/thesis.pdf?sequence=1, accessed 15 January 2015.

Gilder, A & M Mamkeli, 2014, Biofuels in South Africa, *Environment Insight* 5 March 2014, at https://www.ensafrica.com/news/biofuels-in-South-Africa?Id=1339, accessed 15 January 2015.

Girard, P & A Fallot, 2006, Review of existing and emerging technologies for the production of biofuels in developing countries, *Energy for Sustainable Development* 10(2), 92.

Griffin, J, The impact of climate change on South Africa, Climate Energy Institute Post 3 April 2012, at http://www.climateemergencyinstitute.com/cc_s_africa_griffin.html, accessed 15 January 2015.

Hughes, A, M Howells & A Kenny, 2002, *Energy efficiency baseline study*, Capacity Building in Energy Efficiency and Renewable Energy (CABEERE) Report No. 2.3.4 P-54126. Pretoria, Department of Minerals & Energy.

Johnson, S, C Shiffman & D Winokur, 2006, World News, *Sustainable Development Law & Policy*, 76.

Letete, T, 2009, *Multiobjectives Modelling of Biofuel Supply Systems*, Master Thesis, at http://open.uct.ac.za/bitstream/handle/11427/12125/thesis_ebe_2009_letete_t.pdf?sequence=1, accessed 15 January 2015.

Letete, T & H von Blottnitz, 2012, *Biofuels in South Africa: A critical analysis*, at: http://www.erc.uct.ac.za/Research/publications/12Letete-Blottnitz-biofuel_policies.pdf, accessed 15 January 2015.

Lotz-Sisitka, H & P Urquhart, 2014, *South African country study*, South African Regional University Association (SARUA), Volume 2, Report 8, May 2014, at http://www.sarua.org/files/SARUA-Vol2No8-SouthAfrica-Country-Report.pdf, accessed 15 November 2016

Lynd, LR, H von Blottnitz, B Tait, J de Boer, IS Pretorius, K Rumbold & WH van Zyl, 2003, Converting plant biomass to fuels and commodity chemicals in South Africa: A Third Chapter? *South African Journal of Science* 99, 499.

Meyer, EL & KO Odeku, 2009, Climate change, energy, and sustainable development in South Africa: Developing the African continent at the crossroads, *Sustainable Development Law & Policy* 9(2), 49.

Nkomo, J, 2009, Energy security and liquid fuels in South Africa, *Journal of Energy in Southern Africa* 20(1), at http://www.scielo.org.za/pdf/jesa/v20n1/03.pdf, accessed 15 November 2015.

Odeku, KO, 2012, An analysis of renewable energy policy and law promoting socio-economic and sustainable development in rural South Africa, *Journal of Human Ecology* 40(1), 53.

Oliveira, LG & A Hira, 2014, *The evolution and replicability of biofuels in Brazil: Model or exception?*, at https://www.wilsoncenter.org/sites/default/files/de%20Oliveira%20-%20final%20version%20-%20The_Evolution_and_Replicability_of_Biofuels_in_Brazil2.pdf, accessed 15 January 2015.

Peters, D, 2013, Minister of Energy – Budget Vote Speech, at http://www.energy.gov.za/files/media/speeches/2013/2013-BUDGET-VOTE-SPEECH-14May2013.pdf, accessed 15 January 2015.

Pradhan, A & C Mbohwa, 2014, Development of biofuels in South Africa: Challenges and opportunities, *Renewable and Sustainable Energy Reviews* 39, 1089, at http://www.sciencedirect.com/science/article/pii/S1364032114005838, accessed 15 January 2015.

ResearchandMarkets, 2015, *Analyzing the biofuels industry in South Africa 2015*, at http://www.researchandmarkets.com/reports/613688, accessed 15 January 2015.

SAPIA / South African Petroleum Industry Association, 2013, *Annual Report 2013*, at http://www.sapia.co.za/publications/annual-reports.html, accessed 15 January 2015.

Sebitoso, AB & P Pillay, 2008, Renewable energy and the environment in South Africa: A way forward, *Energy Policy* 36, 3312.

Teles da Silva, S & C Dutra, 2014, *Brazilian policy of biodiesel alternative to mitigate effects of climate change*, at https://www.google.de/url?sa=t&rct=j&q=&esrc=s&source=web&cd=1&ved=0ahUKEwjvhvn0w93QAhXExFQKHdTUAa4QFggqMAA&url=http%3A%2F%2Fwww.iucnael.org%2Fen%2Fdocuments%2F99-teles-da-silva-and-dutra-brazilian-biodiesel-policy-mitigation-of-cc%2Ffile&usg=AFQjCNGjO49RGC_j4XsYiiuxJs1V51yaGg&bvm=bv.139782543,d.ZGg, accessed 15 January 2015.

Strydom, EM, 2009, *Biofuel production in South Africa: a critical analysis of the government's biofuel industrial strategy*, Master Thesis, at thesis.eur.nl/pub/6610/Strydom%20Morne,%20Elrich.pdf, accessed 15 January 2015.

Tyler, E, 2011, Aligning South African energy and climate change mitigation policy, *Climate Policy* 10(5), 575, at http://www.tandfonline.com/doi/abs/10.3763/cpol.2010.0094#.VZz-0Pntmko, accessed 15 January 2015.

US Energy Information Administration, 2013, *International Energy Statistics*, at http://www.eia.gov/cfapps/ipdbproject/IEDIndex3.cfm, accessed 15 January 2015.

van Vuuren, RJ, 2008, Alternative energy for Africa: Enhancement of the UN Millennium Development Goals, *Science Scope* June, 54, at http://www.csir.co.za/news/2008_july/pdfs/sciencescope.pfd, accessed 15 January 2015.

van Zyl, E & BA Prior, 2009, *South African biofuels*, IEA Taskgroup 39 Progress Report, at http://academic.sun.ac.za/biofuels/Media%20info/South%20Africa%20Biofuels%20May%202009%20Progress%20Report.pdf, accessed 15 January 2015.

von Maltitz, GP & A Brent, 2008, *Assessing the biofuels options for Southern Africa*, at http://researchspace.csir.co.za/dspace/bitstream/10204/2579/1/Von%20Maltitz_2008.pdf last accessed 15 January 2015.

von Maltitz, G, L Haywood, M Mapako & A Brent, 2009, Analysis of opportunities for biofuel production in sub-Saharan Africa, *Environment Brief*, Center for International Forestry Research (CIFOR), Bogor, Indonesia.

WEC / World Energy Council, 2013, *World Energy Resources*, at http://www.worldenergy.org/wp-content/uploads/2013/09/Complete_WER_2013_Survey.pdf, accessed 25 May 2014.

Winkler, H, 2006, *Energy policies for sustainable development in South Africa: Options for the future*, at http://www.erc.uct.ac.za/Research/publications/06Winkler-Energy%20 policies%20for%20SD.pdf, accessed 15 January 2015.

Regulation of biofuels in South Africa: Most recent developments*

Johann H Krige, BA (Law), LLB, LLM (Stell), Senior Parliamentary Researcher, Democratic Alliance, Cape Town, South Africa

A. Introduction

The technological advances in the production of biofuels augur well for the establishment of a biofuels industry, yet technological capability means little without a supportive regulatory framework. Arguably one of the most prohibitive factors in the production of biofuels in South Africa is the lack of a formal and finalised regulatory framework and the policy uncertainty created thereby. Having a functioning biofuels framework in South Africa will serve multiple governmental purposes, the most promising being job creation and subsequent poverty reduction. However, as will become clear in the discussion below, a supportive legal framework will have considerable fiscal implications for the state purse. As such, both the Department of Energy[1] and the National Treasury[2] has the unenviable task of deciding whether the potential outcomes justify the financial input required. This decision becomes increasingly complex when taking into consideration the political and economic reality in which South Africa finds itself at this time.

While the South African government has been relatively vocal in its agreement with the creation of a supportive legal framework for biofuels, the regulatory process has been in motion for more than a decade without delivering concrete results. The purpose of this chapter is to discuss and analyse the latest developments in the regulation of biofuels in South Africa by referencing the latest documents made available by the Department of Energy during the course of 2015 and 2016. The discussion will be based mostly on official governmental documents; but, as the law does not

---

* This chapter is partially based on the author's LLM thesis conducted under the supervision of Prof Dr Oliver C Ruppel at the Faculty of Law, Stellenbosch Univeristy, South Africa.

1 See http://www.energy.gov.za/home.html for further information.

2 See http://www.treasury.gov.za for further information.

operate in a vacuum, reference will be made to wider policy, political and economic considerations.

There are two scenarios, in the opinion of the author, in which the production of biofuels should be encouraged. The first and most ideal situation would be one in which biofuels can be produced efficiently and in such conditions that government subsidies are either not required or required at a minimal level. As will be seen in the discussion below, however, most predictions indicate that this scenario is not achievable in the current conditions. The second scenario is one in which the benefit to society outweighs the costs to government. This balance of costs is not strictly objectively determinable, and is a matter of government policy rather than law. The aim of this chapter is to discuss and analyse the legal framework in which the government has to develop its policy, rather than to prescribe or make suggestions about policy. However, in the words of Mike Peo, head of infrastructure, energy and telecommunications at Nedbank Capital, the full potential of renewable energy in Africa as a whole will only be unlocked if there is a substantial commitment to the development and implementation of regulatory frameworks that encourage investment in renewable energies.[3]

B. *Renewable energies in South Africa*

I. Liquid fuels in South Africa

The South African economy, still in a developmental stage, is energy-intensive and it is forecast that demand will double by 2030.[4] Owing to the fact that South Africa remains dependent on its energy-intensive coal mining industry, the country accounts for approximately 30% of Africa's total primary energy consumption. As such, South Africa emits the most greenhouse gases among African countries and is responsible for 40% of the continent's emissions, along with being the 13th largest emitter in the world.[5] The issue of massive emissions is exacerbated by the lack of a properly functioning rail network, which means that South Africa is highly dependent on road transport and subsequently the use of liquid fuel.

3 Williams (2016).
4 US Energy Information Adminstration (2015).
5 Based on 2012 estimates, see US Energy Information Administration (2015: 2).

South Africa does not have sufficient proven reserves of oil and natural gas and therefore imports most of its crude oil from Saudi Arabia, Nigeria and Angola. In 2014, South Africa supplemented its local supply of crude oil by importing 425,000 barrels of oil per day.[6] During the later stages of 2015 and early stages of 2016, crude oil prices were at the lowest they have been in a decade. However, political turmoil and a bleak economic outlook contributed to the deterioration of the South African currency, the rand, against most major currencies. The South African consumer has thus not benefited from the low international oil prices. The current situation of importing such a large proportion of liquid fuels is not sustainable and has a severe detrimental effect on the Treasury's balance sheets.

The desire to produce more energy locally is one of the factors that led to the investigation of biofuels as an alternative fuel source.[7] Dependence on foreign sources of liquid fuel clearly impedes the country's energy security, as external influences could easily have an impact on the availability of fuel supply. It further exposes consumers to severe price shocks, an undesirable possibility when taking into consideration the fact that South African consumers are already exposed to high inflation and a subsequent increase in the cost of living.[8]

II. South Africa's developmental goals

South African governmental policy is currently guided by the National Development Plan (NDP), which aims to eliminate poverty and reduce inequality by 2030.[9] The NDP has identified various enabling milestones required to achieve this goal, including an increase in employment and per capita income, and broadening the ownership of assets for historically disadvantaged groups.[10] The NDP also acknowledges the importance of producing "sufficient energy to support industry at competitive prices, ensuring access for poor households, while reducing carbon emissions per unit of power by about one-third".[11] The goals set out by the NDP are to in-

---

6 US Energy Information Adminstration (2015: 2).

7 Pradhan & Mbohwa (2014: 1092).

8 The price for a liter of 95 octane petrol was R12.62 ($0.79) in May 2016, with prices expected to increase by 52 cents in June and more increases in the following months, see *Fin24* (2016).

9 NPC (2013).

10 (Ibid.: 24).

11 (Ibid.).

form any and all policies/regulations created by governmental departments – and the regulation of biofuels is no exception.

In terms of a regulatory framework for biofuels, this would mean that the framework should aim to achieve at least three outcomes. Firstly, the creation of a biofuels industry should reduce poverty through creating employment. Secondly, the industry should develop in such a manner that economic ownership of assets used in the biofuels industry is made accessible to historically disadvantaged groups. The final outcome can be divided into three facets: increasing the power supply, at competitive prices, while reducing carbon emissions per unit of power produced.

C. *Historic development of the biofuels regulatory framework*

I. The Biofuels Task Team

There are clearly various policy considerations that need to be taken into account by the governmental departments, headed by the Department of Energy, involved in creating the final biofuels framework. To align a potential biofuels framework with the NDP, the framework should aim to achieve competitive pricing, poverty reduction, and broadening of ownership. The Renewable Energy Policy White Paper of 2003 sets out a further five areas of facilitation that require consideration (financial instruments; legal instruments; technology development; awareness raising, capacity building and education; governance).[12] When viewing the White Paper as a whole, it becomes clear that the government is aware that a holistic approach to the development of a national strategy for renewables is required to establish a sustainable contribution to the energy mix, which is thus also applicable to biofuels.

In recognition of the complexity of the matter at hand, the South African Cabinet authorised the establishment of a Biofuels Task Team (BTT) in 2005. The BTT consisted of representatives from the departments of agriculture, environment, land affairs, energy, science and technology and the Treasury, along with industry experts. The diversity in the compilation of the BTT is an indication of both the complexity and the holistic approach required for the creation of a biofuels regulatory framework.[13] The BTT was tasked with conducting a feasibility study on biofuels in South

12 GRN (2015a).

13 See http://bit.ly/1Nzfofa, accessed 22 December 2015.

Africa, with due regard to the developmental goals and policy objectives set out by the NDP and the Renewable Energy White Paper.[14] The BTT's findings were presented to Cabinet, and the Biofuels Industrial Strategy (BIS) was subsequently ratified in December 2007.

II. The Biofuels Industrial Strategy

In the BIS, the BTT identified a number of goals that would ideally be achieved through the creation of a biofuels industry. Provision of sustainable income opportunities and subsequent poverty alleviation, substitution of foreign oil imports, attraction of private investment to rural areas and the promotion of agricultural development were the key indicators of success identified by the BTT.[15] To achieve the full potential of the identified goals, the BIS set the ambitious target of achieving a 2% penetration of biofuels in the national fuel supply. The BTT estimated that this level of penetration would lead to the creation of at least 25,000 jobs, mostly in rural farming areas.[16]

Despite the promising results that were envisaged in the BIS, there is still no binding regulatory framework. The delay in implementation is ascribed to political inaction and the very valid concerns about the detrimental effects that the production of biofuels may have on food security.[17] This is despite the fact that the Southern African Development Community (SADC) Biofuel Taskforce approved the SADC Framework for Sustainable Biofuels in December 2009.[18] The Taskforce adopted the Framework with the hope that it would urge member states to "accelerate their initiatives in developing biofuels as a source of alternative and cheap environmentally friendly fuel but also for rural development and poverty reduction."[19] While the SADC Framework is not legally binding on the South African government, it did serve to provide some impetus for the government to accelerate the process which had been stagnant for the better part of four years.

---

14 Pradhan & Mbohwa (2015: 1092).
15 (Ibid.).
16 (Ibid.: 1093).
17 (Ibid.).
18 Available online at http://bit.ly/1ZmbStS, accessed 15 October 2016.
19 SADC (2009: 1).

III. Implementation of the Biofuels Industrial Strategy

One of the major issues identified by the BTT was that biofuels projects would not be viable at prevailing feedstock and crude oil prices, an issue that remains relevant at the time of writing. To address this problem, the Department of Energy attempted to identify measures that would allow it to focus on implementation, monitoring and refinement of the BIS.[20] As a result of these efforts, the Department initiated the Arengo 316 Project. The project, also known as the Cradock Sugar Beet to Bioethanol Project, served as a case study to identify key issues which would need to be addressed by future policy and regulatory, technical and other enabling mechanisms.[21] On initiation, the project's aim was to produce up to 90 million litres of fuel-grade ethanol on an annual basis, while being managed in accordance with the principles set out in the BIS.[22] For it to be deemed a successful experiment, the Arengo Project would thus have to simulate the conditions that would be applicable to a private manufacturer so as to identify the challenges to sustainable manufacturing in the South African market conditions.

IV. Identified challenges to biofuels production in South Africa

1. Viability of manufacturing

Even though information on the exact outcome of the Arengo Project is hard to come by, it did deliver in that it provided the Department of Energy with issues that needed to be addressed. Firstly, it became evident that the biofuels industry would not be able to function profitably if left to free market forces.[23] Therefore the creation of a regulatory framework that provides for the shortcomings of the market is essential for the existence of a captive market. One example of how the government aims to influence market forces is through the Mandatory Blending of Biofuels with Petrol and Diesel Regulations (the "Mandatory Blending Regulations"), which compel manufacturers and wholesalers of petroleum products to buy all biofuels produced by licensed manufacturers and blend them into

---

20 GRN (2014: 9).
21 (Ibid.: 10).
22 Industrial Development Corporation (2013).
23 GRN (2015b).

the national fuel supply to achieve 2% penetration.[24] The Mandatory Blending Regulations will be discussed in more detail below, but their main goal is to provide manufacturers with a guarantee that the entirety of their inventory would be bought up by petroleum manufacturers, and thus mitigate the risk of exposure to market forces.[25]

Other solutions to the challenges identified through the Arengo Project are the development of criteria for the licensing of biofuels manufacturers to ensure compliance with the Petroleum Products Act 120 of 1977; the development of an appropriate Biofuels Pricing Framework to regulate the biofuels transfer price between manufacturers of biofuels and the petroleum companies responsible for blending biofuels into the fuel supply; and the establishment of a financial support scheme for qualifying manufacturers that would contribute to the financial viability of large-scale projects.[26]

2. Benchmark feedstock

External parties, in the form of observers and industry stakeholders, further identified three prominent issues relevant to the establishment of the biofuels industry in South Africa. The first issue is that the government's strategy failed to adequately address the issue of benchmark feedstock chosen for biofuels production.[27] The crop chosen as benchmark feedstock would have an impact on the pricing framework on which the government's financial subsidies would be based. In making the final decision on which crop would be used as benchmark feedstock, food security and the availability of water resources would have to be considered. For example, maize was excluded as a potential ethanol feedstock, owing to it being a major staple crop and source of animal feed, despite the increase in maize production brought about by technological advances. South Africa is, in fact, capable of producing a surplus of maize with the right incentives. It is understandable that maize farmers are dissatisfied with the decision to exclude maize as a feedstock, as this has the consequence of denying them a possible income from biofuel production.[28]

---

24 Government Notice No. R.719 (30 September 2013) of Government Gazette No. 36890.

25 GRN (2014: 10).

26 GRN (2015b).

27 Pradhan & Mbohwa (2015: 1093).

28 (Ibid.).

3. Food versus fuel

Sub-Saharan Africa lacks food security, and concerns have been raised about the impact that biofuels production may have on both the availability and price of food. According to one study, prices could increase between as little as 3% and as much as 30%.[29] A conflicting study concluded that biofuel manufacturing would not adversely affect food security in South Africa as production would compete with feedstock destined for export rather than feedstock destined for local consumption.[30] A less obvious consequence is that, owing to the emphasis of BIS on rural development, food security may decrease on a more local scale. A large portion of the rural population is involved in subsistence farming, but may be tempted rather to seek employment at major biofuel projects.[31] This argument too has a counter-argument: as the income level of the rural population grows, so does their purchasing power, which allows them to buy food rather than being dependent on subsistence farming.[32]

Proponents of biofuels may argue that approximately 14% of available arable land in South Africa is being underutilised, and that only 1.4% of that land would be required to grow the crops necessary to achieve the 2% penetration level envisaged by the Mandatory Blending Regulations.[33] However, the recent drought experienced in South Africa highlighted shortcomings in South African disaster mitigation capabilities, exacerbated by poor infrastructure and lack of sustainable governance by the Department of Water Affairs and Sanitation.[34] There is no clear winner in the food versus fuel debate, but it is clear that there are a multitude of considerations and unforeseeable events that will continue to inform the debate. Regardless of the outcome of the debate, if there ever is one, it is imperative that an efficient and sustainable biofuels strategy must take into account that food security is likely to be influenced by changing weather patterns as the effects of global warming become more apparent.

---

29 Gasparatos et al. (2015: 889).
30 Arndt et al. (2010).
31 Gasparatos et al. (2015: 889).
32 (Ibid.: 890).
33 Pradhan & Mbohwa (2015: 1094).
34 Bothma (2015).

4. Social implications

As the manufacturing process of biofuels is such a complex one, the consequences of manufacture are not necessarily clear at the outset of manufacturing and can be wide-reaching. While most of the consequences are seen to improve health, safety, human rights and livelihood, any regulatory framework should guard against abuses. In South Africa, the BIS focuses on the development of biofuels in rural areas, where the legacy of an oppressive Apartheid regime remains and leaves the rural population vulnerable to abuses of power. A major concern is that the developers of biofuels projects may use confusing tactics to trick rural land owners to part with their land at below-market prices, causing damage to the livelihood of rural communities.[35]

There are various other problems that are unique to the South African context. Large numbers of illegal migrant workers from South Africa's neighbouring countries, low wages, crimes against commercial farmers and child labour are notable concerns that may affect the expansion of biofuel manufacturing in South Africa.[36] It is evident that there is thus a multitude of factors that require consideration when embarking on a major biofuels project, and it is essential that the regulatory framework provides protection against the possible negative side-effects that could occur.

D. The Draft Position Paper on the South African Biofuels Regulatory Framework

I. Introduction

The Department of Energy published the Draft Position Paper on the South African Biofuels Regulatory Framework ("Position Paper") in January 2014. The intention of the publication of the Position Paper was to start the process of developing and finalising a regulatory framework for the production of biofuels in South Africa. In the Position Paper, the Department of Energy acknowledges the failure of the BIS to lead to an effective biofuels industry, but declares that it remains committed to overcoming the identified challenges with the cooperation of relevant stake-

35 German et al. (2011).
36 Mbohwa & Myaka (2010).

holders and actors within the industry.[37] In this section, the most important aspects of the Position Paper, as it relates to the development of the current regulatory framework, will be elucidated and discussed.

II. Licensing of manufacturers

Since a major part of the government's reasoning behind the promotion of the biofuels project is the achievement of socioeconomic goals, it was important that criteria were developed to ensure that potential manufacturers would contribute to the achievement of that goal. As such, the Department of Energy developed the comprehensive Criteria for Licensing Manufacturers of Biofuels ("Criteria"). The Criteria were developed in terms of the Petroleum Products Act 120 of 1997, although they also incorporate the goals set out in the BIS. The Criteria further take into account the sustainability issues prevalent in South Africa in the context of the aforementioned Act. The Office of the Controller of Petroleum Products in the Department of Energy is responsible for evaluating all applications by potential manufacturers of biofuels, and to decide whether to grant such a licence.[38]

Before the implementation of the Biofuels Draft Regulatory Framework, which will be discussed below, the criteria used for licensing was also used to determine whether manufacturers would qualify for financial support from the Department. Criteria included contribution to the transformation of the liquid fuels industry, facilitation of social inclusion, protection of agricultural land rights, positive contribution to rural development, avoidance of food security threats, protection of scarce natural resources and the ability to produce biofuels sustainably.[39] As things stand, eight manufacturers have received manufacturing licences. Four companies, namely Mabele Fuels, Ubuhle Renewable Energy, E10 Petroleum Africa and Arengo 316, are bioethanol manufacturers, and have a total combined manufacturing capacity of 392.2 million litres per annum. A further four biodiesel manufacturers have received licences, with a total combined blending capacity of 970 million litres per annum, namely Rainbow Nation Renewable Fuels, Exol Oil Refinery, Phyto Energy and

---

37 GRN (2014: 1).

38 GRN (2014: 12).

39 For more information on the exact criteria, see GRN (2014: 26-27).

Basfour 3582.[40] At the time of publishing, there was no publicly available information that any of these manufacturers are manufacturing biofuel to be used in the South African liquid fuels pool.

III. Biofuels pricing Framework

The Biofuels Pricing Framework (BPF) was developed to facilitate investment in the biofuels industry by providing a guarantee to investors that they would receive sufficient returns on their investments. The government hoped that this would attract investors to the industry. The BFP was developed to minimise the costs involved with manufacturing to determine the break-even point of manufacturing costs and to establish the level of subsidies to be granted to manufacturing entities.[41]

To determine the levels of subsidies payable to a manufacturer, the Department of Energy launched the Biofuels Break-even Price Determination Project that was to determine the break-even price for an optimally sized and efficiently operated biofuel manufacturing plant with a rate of return on assets of 15%. The Bio-ethanol Blending Value Project was tasked with determining the true value of blending bioethanol with conventional liquid petrol and supplying the blend to end-users of petrol.[42] However, it became clear that the system of subsidies proposed in terms of the BFP would not be financially sustainable for the state, and a new approach was adopted in the Draft Regulatory Framework, which will be discussed in the next section.

IV. Reference crops and subsidies

During the development of its policy, the Department of Energy conducted studies into the viability of different crops for use in the manufacture of biofuels. The assumption was that sorghum would be used for the manufacture of bioethanol and soya beans for the production of biodiesel. It comes as no surprise that manufacturers that planned on using feedstock other than sorghum or soya beans (the two reference crops) were not satisfied with the assumptions used in the pricing model. If the Pricing

---

40 (Ibid.).
41 (Ibid.: 12).
42 (Ibid.: 14).

Framework were to be based on only those two reference crops, it would effectively disincentivise manufacturers that were planning on using other feedstock crops from investing in the industry, thus placing the viability of the project in jeopardy.[43] The issue of which reference crops to use in the pricing model caused a major delay in the development and adoption of the Pricing Framework, as the BTT had to consider alternatives to the proposed plan of using the reference crops without exposing the state purse to undue pressure.[44]

The selection of sorghum and soya beans as reference crops was not arbitrary, but rather the result of the Bio-Ethanol Value Determination and the Determination of the Break-even Price of Biofuels Manufacturing projects. However, the studies found that sugarcane and sunflower seeds were also viable crops for the production of bioethanol and biodiesel, respectively. All four of these crops, namely sorghum, soya beans, sugarcane and sunflower seeds, are already grown in large quantities in South Africa, and the industry has the experience to expand the production of these crops. Sugar beet was also considered as feedstock crop.[45]

The BTT subsequently considered three options in determining the reference crop, in an attempt to attract the highest number of investors at the lowest cost to the taxpayer. The first option was to consider only one reference crop, or the most efficient reference crop. The reasoning behind this option was that it would be in the best interest of those burdened with paying the increased fuel levy, as the Pricing Framework would only cover the difference in income between actual income and the income required to achieve the 15% return of assets for the most efficient method of producing biofuels. Additionally, this option would promote financial sustainability, improve transparency, and lead to certainty with regard to the calculation of the subsidy.[46]

The second option would be the consideration of multiple reference crops in the pricing model. This would entail using sorghum as the reference crop for grains and starches and sugarcane as the reference crop for sugar-based crops for the production of bioethanol. However, this option would still only consider soya beans as the reference crop for biodiesel, as the difference in financial viability between soya beans and other crops is too big.[47] The final option would be a crop-neutral bidding process, by ra-

43 (Ibid.: 18).
44 (Ibid.).
45 (Ibid.).
46 (Ibid.: 20).
47 (Ibid.).

ther soliciting bids from prospective manufacturers. The bids would be received in terms of the Renewable Energy Independent Power Producer Procurement Programme (REIPPPP). In order to submit a bid, the potential manufacturer would first have to qualify for licensing by meeting the relevant criteria, which will be discussed below. The bidders would indicate the amount of subsidy needed to achieve the 15% return on assets at different pricing levels. The BTT found that the bidding model would be too complicated to adopt, at least in the initial phase, without actually guaranteeing a superior end-product.[48]

In the end, the BTT recommended that only sorghum and soya beans be considered as reference feedstock.[49] The BTT argued that this approach would serve to expedite implementation of the BIS. However, a manufacturers' decisions to use crops other than the reference crop would not disqualify them from receiving the subsidy. If prospective manufacturers satisfy the criteria to qualify as beneficiaries, they would be able to apply for the subsidy.[50] The Draft Position emphasises that this approach is only to be followed in Phase One of implementation, and that further deliberation is needed to establish the best way to determine the subsidy.[51] It has now become clear that the pricing scheme adopted by the BTT was not successful in facilitating the growth of a biofuels industry in South Africa. In consequence, a different approach of providing subsidies has been suggested in the Draft Regulatory Framework, which follows an approach similar to the final suggestion discussed above.

V. Problem areas

The Draft Position Paper was developed with the understanding that government would use it as an opportunity to learn lessons from failures and successes identified through engagement with industry stakeholders. The biofuels projects were also proposed against the backdrop of a weakening local and global economy, intensifying weather pattern changes, and an ever-increasing level of unemployment in South Africa. It is clear that the requirements placed on potential manufacturers would increase manufacturing costs, and therefore trigger the need for a pricing mechanism in

48 (Ibid.: 23-24).
49 (Ibid.: 24).
50 (Ibid.).
51 (Ibid.).

which government subsidises the increase in costs. The Draft Regulatory Framework provides a previously absent and much needed semblance of policy certainty. As will be discussed below, manufacturers now have an idea of how they will have to go about receiving the subsidy from government, which is essential for the financial viability of their projects.

E. *The Draft South African Biofuels Regulatory Framework*

I. Introduction

The most recent public information on the status of the government's biofuels project is found in the National Treasury's Estimates of National Expenditure of the 2016 National Budget.[52] The document simply states that the Department of Energy is currently in the process of reviewing the proposed biofuels subsidy models, as well as further investigating the concerns about food security. Further official information on the status of the biofuels project is absent, and it is not clear whether any entity is currently manufacturing biofuels on a large scale.[53]

The Draft South African Biofuels Regulatory Framework ("Draft Framework") was presented by the Department of Energy in June 2015, but the document was not yet publicly available at the time of publication. However, this document is the clearest and most up-to-date indication of how the government intends to regulate biofuels in the years to come. The Draft Framework is illustrative of the responsiveness of the Department of Energy to the issues identified by stakeholders since the Draft Position Paper was published, although it does not address all possible issues.

The Draft Framework focuses on four issues identified as the main areas that require regulation to ensure the success of the biofuels industry in South Africa. The Framework provides for the mandatory blending of biofuels into liquid transport fuels in order to ensure demand for biofuels. The Biofuels Pricing Framework is further refined in the Draft Framework to create certainty with regard to the pricing mechanism for the sale of biofuels to large-scale fuel manufacturers. The greatest deviation from the Draft Position Paper comes in the form of the biofuels subsidy mecha-

52 National Treasury (2016).

53 On the contrary, one of the potential biofuels manufacturers, PhytoEnergy, has reverted the focus of its business to export rather than providing to the local market, until such a time when supplying the local market becomes profitable. See http://goo.gl/fFqOMm for more details.

nism. Rather than following a uniform approach, biofuels manufacturers will now have to apply for subsidies and subsidies will awarded through a transparent process to those manufacturers that meet the socioeconomic indicators identified by the government, while producing biofuels at the lowest cost reasonable in the circumstances. Finally, the framework provides for the mechanism that will regulate the cost recovery mechanism for the blenders (large-scale fuel manufacturers, in most cases) to ensure that the cost of blending remains cost neutral for the blenders.[54]

The Draft Framework further states that the effects of implementation of the programme will be monitored and reported to Cabinet, which will then be tasked with determining the next phase of implementation.[55] The most important observation that needs to be made is whether 2% penetration is actually achievable, and, if so, what the socioeconomic benefits, costs to the economy and issues arising from the industry are. To minimise the risk to the government, the initial phase of the project may be implemented in phases.[56]

II. Mandatory Blending Regulations

The Mandatory Blending Regulations came into effect on 1 October 2015. The reasoning behind the Mandatory Blending Regulations is that it would create a captive market for biofuels by requiring that a level of at least 2% of bioethanol and 5% of biodiesel be blended into petrol and diesel respectively.[57] The date set by the minister was chosen with the understanding that all outstanding issues would have been resolved by that date and that manufacturers would be able to produce the necessary amounts of biofuels to meet the blending regulations.

In an attempt to ensure compliance, the minister of Energy approved the establishment of the Biofuels Implementation Committee (BIC). The BIC was to be chaired by the Department of Energy and consisted of actors from the oil and fuel industries as well as representatives from Transnet, the state-owned transport authority, when the situation called for it. The BIC was assigned the goal of addressing all matters pertaining to the practicalities of blending biofuels with petrol and diesel. The BIC enjoys

54 GRN (2015b: 4).
55 (Ibid.).
56 (Ibid.: 5).
57 (Ibid.: 7).

the support of the Biofuels Task Team, as mandated by Cabinet.[58] The decision to implement blending regulations is not only based on incentivising manufacturers with a guarantee that their stock will be bought up. It also forms part of the Cleaner Fuels Programme (CF), promulgated by the Department of Energy. CF 1 was first promulgated in 2006, and was followed up by CF 2, adopted in 2012.[59] One of the main objectives of CF 2 was to reduce the sulphur levels in South Africa's fuel supplies to 10 parts per million, and thus in line with Euro V-specification levels. The Department of Energy set July 2017 as the target date for reaching that goal, but has since abandoned that date and is yet to set a new target date.[60]

At the time of writing, no large-scale producer of biofuels was yet operational, thus the October 2015 deadline of the Department of Energy was missed. Until such time as there is regulatory certainty surrounding biofuels, it is unlikely that any manufacturer will start construction on a major biofuels project. In hindsight, the date chosen by the Department in 2013 was probably unrealistic, as there had been no development in terms of biofuels for over six years when the deadline was set. The Department naively believed that the process could be expedited and that several biofuels manufacturers could be online within two years of setting the target date, despite evidence that there were still many outstanding issues that needed to be resolved. This displays an underlying shortcoming in the approach of the Department of Energy to policy, namely setting ambitious targets without providing sufficient support for the initiatives required to achieve those targets.

III. Biofuels Pricing Framework

Soon after the publication of the Draft Position Paper, it became apparent that the pricing mechanism proposed by the Department of Energy would not be affordable in the current economic climate. The first mention of a new mechanism was brought to the attention of the public in August 2015. Under the new proposed incentive, manufacturers would have to compete directly with each other, adding a competitive element to receiving the subsidy. While such an approach makes more sense in terms of affordabil-

58 GRN (2014: 11).
59 (Ibid.: 12).
60 Creamer (2014).

ity of the project, manufacturers fear that this might drive down investor confidence, as a consequence of an increase in investment risk.[61]

Regardless of the fears raised by manufacturers, the government did indeed adopt a mechanism whereby manufacturers will have to bid for delivered prices to the BTT, which will then have to select projects that will receive subsidies from the government. The reasoning behind this approach is that it will apparently expose those involved to the risks related to feedstock price, as manufacturers will be aware of capital costs and fixed operating costs at the time of submitting their bid.[62]

The biofuels bid price will subsequently consist of two parts: the biofuels conversion price and the feedstock price. The biofuels conversion price reflects the cost of recovery of capital expenditure, fixed operational costs and required return on investment. The feedstock price is less certain than the biofuels conversion price, as it requires an estimate of the costs involved in securing the feedstock required to produce biofuels. The bid price will be calculated on the basis of the rand cost per litre. Costs will be escalated at a basket of indices that reflect the main input costs for the duration of the agreement.[63]

Those entities responsible for blending the biofuels into the conventional fuel supply will then be responsible for buying the biofuels from the manufacturers at the biofuels bid price, and the costs incurred by the blender will be covered by the biofuels subsidy. The envisaged objective is to guarantee cost-neutrality for the blenders.[64]

IV. Cost recovery mechanism for blending costs

Contamination is a serious risk when biofuels are transported, which means that blending can only take place at petroleum depots. The petroleum companies responsible for the blending of biofuels into the fuel supply will have to make capital expenditures to facilitate blending, and there are operational costs involved with blending. The blenders will be able to recover these costs from the government, along with the additional cost of procuring biofuels instead of mineral fuel, through the application of the

61 Roelf (2015).
62 GRN (2015b: 7).
63 (Ibid.: 8).
64 (Ibid.).

Regulatory Accounts System or another mechanism that may be more appropriate under the circumstances.[65]

V. Biofuels Feedstock Supply Protocol

The possible impact of a biofuels industry on food security in a country like South Africa that faces regular droughts and other challenges to food production is an issue that should be at the forefront of any policy developer's mind. The Department of Energy acknowledges the danger of food security being adversely impacted by the manufacture of biofuels, and as such has collaborated with the Department of Agriculture, Forestry and Fisheries to develop the Biofuels Feedstock Supply Protocol ("Feedstock Protocol"). The Feedstock Protocol is implemented to achieve a balance between the BIS objectives of job and wealth creation and the possible negative effects that large-scale biofuels manufacture may have on food security and the environment. The Feedstock Protocol attempts to do so by focusing on three indicators that are deemed to be crucial to the success of the project, namely mitigation of risk to food security, reduction of impact on the environment, and provision of support to the domestic agricultural sector and emerging black farmers.[66]

1. Mitigation of risk to food security

As the South African biofuels strategy is, at present, concerned with first generation biofuels, it is essential that safeguards are put in place to ensure that both the availability and affordability of food are impacted to the least possible extent. The Biofuels Feedstock Protocol does this by prioritising the use of crop-free land in the production of biofuels feedstock, prohibiting the use of maize as a feedstock, prioritising production that does not require irrigation, and only permitting biofuels production on farms where feedstock production is part of plant rotation that boosts the financial viability of the farm.[67] Whether these measures will be sufficient to negate the adverse effects of biofuels production on food security will only become clear with time.

---

65 (Ibid.: 9).
66 (Ibid.).
67 (Ibid.: 10).

The final determination that only farms that produce feedstock to augment the financial viability of the farms could be deemed to be controversial, as it places a restriction on new farmers that might want to enter the biofuels market through providing feedstock to biofuels manufacturers. However, the reasoning behind the provision must be that this provision will prevent farmers that are in the business of producing food crops converting their operation to the production of biofuels feedstock. If too many farmers were to make such a switch, food security will surely be adversely impacted. The Department, however, should allow for exceptions to the rule where circumstances dictate that unused or underutilised land can be converted to feedstock production without negatively impacting food security.

2. Reducing impact on the environment

Mitigating risk to food security cannot be achieved through only limiting the feedstock or land that may be used for biofuels manufacture. It also requires that long-term protection is granted to water supplies and the country's biodiversity. As such, the use of jatropha, a popular feedstock for biofuels manufacture in other parts of the world, is excluded, as it may have adverse effects on biodiversity. The Feedstock Protocol also promotes the use of rain-fed feedstock production. A water use licence may be granted by the Department of Water and Sanitation in exceptional circumstances.[68] By protecting water resources and biodiversity, the likelihood that resources that could be used in food production are diverted to the production of biofuels is decreased.

3. Supporting the domestic agricultural sector and emerging black farmers

One of the core principles of the BIS is that a reasonable portion of the benefits and profits accruing from the manufacture of biofuels vests in rural black farmers. This not only adds a legitimate government purpose to the project, but could also contribute to eradicating the legacy of economic inequality imposed by Apartheid legislation, if implemented correctly. To

68 (Ibid.). The water use licence is regulated by Section 21 of the National Water Act 36 of 1998.

advance the agenda of creating wealth for previously disadvantaged communities, the Feedstock Protocol requires biofuel manufacturers to source the entirety of their feedstock from local farmers within three years of qualifying for the biofuels subsidy. Importation of feedstock will only be permissible in the event of crop failure due to natural disaster.[69]

VI. Biofuels manufacturer selection process

It goes without saying that only a select few large-scale manufacturers of biofuels can qualify for the subsidy without placing too large a strain on the fiscus. The volume cap for Phase One of implementation of the BIS remains at 2% of the national fuel demand in the Draft Regulatory Framework, which is estimated to be the equivalent of 460 million litres of biofuels per annum. As a further precaution against unforeseen risks related to the subsidy, the Draft Regulatory Framework states that Phase One may be introduced in stages.[70] As the manufacturers will play a role in achieving governmental goals, the subsidy will be rewarded to manufacturers in accordance with the principle of value for money – that is the quality and number of jobs created in relation to the amount of subsidy received – through a fair and transparent selection process. Once a manufacturer has been selected for the subsidy, an Implementation Agreement will be concluded with the government.[71]

In order for a prospective manufacturer to qualify for the subsidy, it will have to fulfil the requirements of a three-stage selection process. The first step will be to comply with a Request for Information issued by the Department of Energy, which will be used by the Department and the BTT to further develop the biofuels regulatory framework in line with the needs of the industry. Once the first step has been completed, the BTT will issue a Request for Proposal to the prospective manufacturers that have complied with the Request for Information. The BTT will look at criteria such as compliance with black economic empowerment codes, readiness for implementation of the project, and the sustainability for production based on the feedstock to be used by the manufacturer. The following table elaborates on the criteria set out in the Draft Framework.

69 GRN (2015b: 10).
70 (Ibid.: 11).
71 (Ibid.).

Table 1: Subsidy qualification criteria[72]

Prequalification criteria	Indicator
1. Compliance with broad-based black economic empowerment objectives of the Biofuels Industrial Strategy	The project should have a minimum of 40% black shareholding where black shareholding is defined as per BBBEE codes.
2. Permitting, Licensing and authorisations	Projects that will be considered should show evidence of: • Biofuels manufacturing licence; • Environmental authorisation and licences; • Water-use licence; • Land-use licence; • Access to land for the biofuels manufacturing plant; and • Ability to raise funding as evidenced by a funding term sheet from a reputable financing institution.
3. Sustainable production of feedstock	A biofuels manufacturer will be required to provide evidence that the Feedstock Supply Plan has been approved by Department of Agriculture, Forestry and Fisheries (DAFF). DAFF will assess the Feedstock Supply Plan for approval against the Biofuels Feedstock Protocol.

If the prospective manufacturer complies with the minimum criteria above, the manufacturer will qualify for consideration for the subsidy. The BTT has developed further adjudication criteria for the unlikely scenario that the combined manufacturing capabilities of the qualifying manufacturers exceed the level required to achieve 2% penetration. The criteria are designed to ensure that the manufacturers that contribute the most to the socioeconomic goals of the project at the most competitive price will receive the subsidy. The criteria include the biofuels bid price, the lead time to biofuels production, the percentage of local spend of the project, additional participation of black shareholding above the required 40%, rural

72 Source: GRN (2015b: 15).

development initiatives and small farmer support, the proportionate number of jobs created, and the level of support provided to small, medium and micro-sized enterprises and the local community.[73]

VII. Implementation process

Owing to the complexity and the ambitious goals of the South African biofuels project, various governmental departments and agencies are involved in the implementation process. The Draft Regulatory Framework has set the timeline for achieving the first two phases of the project, namely the preparation and project selection phases, at 18 months, presumably from the presentation of the Draft Regulatory Framework (i.e. June 2015). Unfortunately, requests for information from the Department of Energy on the progress of the project were not answered by the time of publication. It is thus unclear what the current state of compliance with the timeline is. It seems, however, that the government intends that the framework required for manufacture be in place after those 18 months. The likelihood that biofuels will be produced on a mass scale any time before 2018 thus seems unlikely.

Table 2: Governmental departments and agencies involved in the implementation process and their roles[74]

National Energy Regulator (Nersa)	Revoking of licences for petroleum storage facilities that become manufacturing facilities in accordance with the Petroleum Pipelines Act 60 of 2003
Department of Energy	Consideration of licensing and permit applications for the manufacturing of biofuels under the Petroleum Products Act 120 of 1977; administration of the subsidy scheme; monitoring and verification of biofuel manufacturing, supply and marketing set out in the Implementation Agreements
Department of Water and Sanitation	Consideration and determination of applications for water-use entitlement for irrigation and industrial use
Department of Environmental Affairs	Consideration and determination of environmental authorisations, licences and permits

73 (Ibid.).
74 (Ibid.: 17).

Department of Agriculture, Forestry and Fisheries	Evaluation and approval of prospective biofuels manufacturers' Feedstock Supply Plan; development and support of black emerging commercial farmers
Department of Science and Technology	Support of research (throughout the biofuels value chain) and investigation of continuous improvement opportunities
National Treasury	Adjustment of the fuel levy to include the required amount to fund the biofuels subsidy
Department of Trade and Industry	Monitoring and verification of socioeconomic benefits set out in the Implementation Agreements

The Annual Performance Plan of the Department of Energy for 2016/2017 provides some insight into an expected timeline for the biofuels project to be actioned.[75] The Department's goal for the 2016/2017 financial year is to achieve ministerial approval for the draft Biofuels Regulatory Framework discussed above, the findings of which will be implemented in the 2017/2018 and 2018/2019 financial years. These goals form part of the Department's five-year Strategic Plan Target for introducing biofuels into the liquid fuels industry as a means of developing liquid fuels energy security.

More specifically, the Department identified the third quarter of the 2016 financial year as the completion date for the revised Draft Biofuels Framework, which will then be submitted for ministerial approval in the final quarter.

F. Conclusion

The fact that a concrete Draft South African Biofuels Regulatory Framework is at least in development is an encouraging sign. However, the timeline set by the Department of Energy indicates that it will take many years before South Africans can realistically expect to have biofuels as part of the national liquid fuel supply. While the Draft Regulatory Framework is a step forward from the Draft Position Paper in terms of providing regulatory certainty, there are various issues that will only become clear with time. For example, the extent of the fiscal impact of the subsidies paid to manufacturers is not clear, nor what the response of the fiscus would be in the case of an economic recession. A clear contingency plan for risk manage-

75 Available online at http://goo.gl/8GSra1, accessed 15 October 2016.

ment is absent in the Regulatory Framework and is something that the Department should seriously consider including.

What remains clear is that a biofuels industry in South Africa does have the potential to address governmental goals such as job creation, greener transport fuels and wealth redistribution, but none of these goals will be achieved until such time as a regulatory framework is finalised and implemented. Recent technological advances have proved that the manufacture of biofuels will become less expensive in the future, but for investors to invest in the further development of the technology there needs to be at least the promise of a lucrative biofuels industry. At present, the risk involved in the manufacture is too high to attract large-scale investment, and the government cannot hope to fund the project without assistance from the private sector.

Bibliography

Arndt, C, R Benfica, F Tarp, J Thurlow & R Uaiene, 2010, Biofuels, poverty and growth: A computable general equilibrium analysis of Mozambique, IFPRI Discussion Paper 00803, at https://www.google.com/url?sa=t&rct=j&q=&esrc=s&source=web&cd=2&cad=rja&uact=8&ved=0ahUKEwjnvOSVkeTPAhXEbhQKHSU1Ba8QFggrMAE&url=http%3A%2F%2Fwww.iese.ac.mz%2Flib%2Fpublication%2FII_conf%2FGrupoI%2FBiofuels_CHANNING%2520et%2520al.pdf&usg=AFQjCNEedkoCvTeHdj9x5Iq0FV7JSTl5FA&sig2=n6zpsFkOlhAKNm6ikXqcbw, accessed 15 October 2016.

Bothma, B, 2015, Recent rainfall won't break drought – experts, *ENCA Online,* 17 November 2015, at http://bit.ly/1mz51iK, accessed 10 October 2016.

Creamer, T, 2014, Mid-2017 clean fuels compliance deadline 'off the table', *Engineering News*, 6 May 2014, at http://bit.ly/1JYLfCd, accessed 30 December 2015.

Fin24, 2016, Huge hike in petrol, diesel prices, at http://goo.gl/yaXkTQ, accessed 30 May 2016.

Gasparatos, A, GP von Maltitz, FX Johnson, L Lee, M Matahi, JA Puppim de Oliveiera & KJ Willis, 2015, Biofuels in sub-Sahara Africa: Drivers, impacts and priority policy areas, *Renewable and Sustainable Energy Reviews* 45, 879.

German, L, GC Schoneveld & P Paheco, 2011, Local social and environmental impacts of biofuels: Global comparative assessment and implications for governance, *Ecology and Society* 16(4), 29.

GRN / Government of the Republic of South Africa, Department of Energy, 2014, Draft position paper on the South African biofuels regulatory framework, at http://www.energy.gov.za/files/policies/Draft_position_paper_on_the_SA_Biofuels_Reg_Frmwrk.pdf, accessed 15 October 2015.

GRN / Government of the Republic of South Africa, Department of Energy, 2015a, State of renewable Energy in South Africa, at http://bit.ly/22fUSrL, accessed 15 October 2016.

GRN / Government of the Republic of South Africa, Department of Energy, 2015b, Draft South African Biofuels Regulatory Framework: The First Phase of the Biofuels Industrial Strategy – Version 5, 19 June 2015, document provided by the Department of Energy on request.

Industrial Development Corporation, 2013, Arengo 316 (PTY) LTD – Cradock Ethanol Project, at http://bit.ly/1YsNytt, accessed 22 December 2015.

Mbohwa, C & N Myaka, 2010, Social life cycle assessment of biodiesel in South Africa: An initial assessment, *Proceedings of the ninth international conference on ecobalance, towards and beyond 2020*, at http://bit.ly/1mziEhQ, accessed 10 October 2016.

National Treasury, 2016, 2016 estimates of national expenditure, 24 February 2016, at http://goo.gl/WfimYz, accessed 15 October 2016.

NPC / National Planning Commission, 2013, National development plan 2030: Our future – make it work.

Pradhan, A & C Mbohwa, 2014, Development of biofuels in South Africa: Challenges and opportunities, *Renewable and Sustainable Energy Reviews* 39, 1089.

Roelf, W, 2015, Cheaper oil forces SA to rework biofuels subsidy, *Reuters Africa,* 11 August 2015, at http://af.reuters.com/article/topNews/idAFKCN0QG0RJ20150811, accessed 15 October 2016.

SADC / Southern African Development Community, 2009, Framework for sustainable biofuels, at http://www.energy.gov.za/files/policies/petroleum/Annexure-D-SADC-Framework-for-Sustainable-Biofuels-Dec-2009.pdf, accessed 15 October 2016.

US Energy Information Administration, 2015, South Africa: International energy data and analysis (29 April 2015), at http://1.usa.gov/1lR2f7G, accessed 10 December 2015.

Williams, F, 2016, Alternatiewe kragoplossings: Afrika moet anders dink oor energie, *Die Burger*, 31 May 2016, 16.

A critical analysis of the regulatory framework of the South African biofuels industry and its compatibility with WTO law

Chengetai E Hamadziripi, LLB Hons, LLM International Trade Law, Lecturer of Law, Midlands State University, Zimbabwe

A. Introduction

Biofuels have been seen as the panacea to a wide range of challenges faced by numerous governments, ranging from environmental to energy security threats.[1] Biofuels have been defined as "liquid, gaseous or solid hydro-carbon fuels derived from biomass".[2] The American Biomass Research and Development Act of 2000 (American Biomass Act) section 303(3) defines biomass as follows:

> [A]ny organic matter that is available on a renewable or recurring basis, including agricultural crops and trees, wood and wood wastes and residues, plants (including aquatic plants), grasses, residues, fibres, and animal wastes, municipal wastes, and other waste materials

Biofuels are considered a renewable source of energy as the substances from which they are derived may be replenished in a short period of time. Biofuels have been classified into 'first generation' and 'second generation' biofuels.[3] First generation biofuels are created from food crops, whereas second generation biofuels are created from lignin, cellulose and hemicellulose, which is derived from the stems, branches and barks of plants.[4] Second generation biofuels are more expensive to produce. Presently, the main focus of commercial biofuels production is on first generation biofuels.[5]

Biofuels can be used as an alternative to, or blended with, transport fuels. The most common form of biofuels available on the market is bioethanol and biodiesel. Bioethanol may be mixed with petrol, and biodiesel

1 Harmer (2009: xi).
2 Howse et al. (2006: 6).
3 (Ibid.).
4 (Ibid.).
5 (Ibid.).

may be mixed with diesel to reduce fossil fuel consumption.[6] Bioethanol is a liquid fuel.[7] It is mostly created from crops which contain starch, such as sugar cane, sorghum, sugar beet and maize.[8] Biodiesel can be produced from vegetable oil or animal fat, and is a liquid fuel that consists of long-chain alkyl esters.[9] Biodiesel is distinct and different from vegetable and waste oils. These waste oils have been used to fuel converted diesel engines.[10] The use of biofuels as an alternative fuel is not new on the energy landscape. Bioethanol was blended with petrol and used as transport fuel at the turn of the twentieth century in America and in Europe.[11] However, interest in the use of biofuels did not increase as large reserves of petroleum oil were discovered.[12]

The first large-scale national production of biofuels began in November 1975 in Brazil, with the establishment of the country's National Alcohol Programme.[13] Other countries have now also implemented policies to encourage the production and use of biofuels.[14] Presently, most biofuels are produced and consumed in domestic markets.[15] A global biofuels industry has the potential to emerge in response to the increasing interest in the use of biofuels. The renewed interest of governments in integrating biofuels as an alternative transport fuel has the potential to increase the trade of biofuels across borders.[16] However, the global trade of biofuels has been limited, with only an estimated 10% of global biofuels production traded across borders.[17] The low level of trade has been attributed to the fact that most countries impose trade-restricting measures to protect domestic biofuels production.[18]

The increase in the international trade of biofuels has been restricted by a number of barriers, the most formidable being the subsidisation of biofuels production. Other significant trade barriers include, *inter alia*, tech-

---

6 Chakauya et al. (2009: 176).
7 Switzer (2012: 250).
8 (Ibid.).
9 Junginger et al. (2009: 2030).
10 (Ibid.).
11 Rosillo-Calle & (2010: 1).
12 (Ibid.).
13 (Ibid.).
14 Switzer (2012: 251).
15 Howse et al. (2006: 4).
16 (Ibid.).
17 Harmer (2009: para. 1.4). The author notes that "despite the enthusiasm with which biofuels have been embraced, global production and trade is relatively small."
18 Harmer (2009: para. 1.4).

nical standards; sustainability criteria and certification standards for biofuels; and sanitary and phytosanitary measures which may be applicable to biofuels.[19]

Establishing a biofuels industry in a country requires government support.[20] This is because biofuels are more costly to produce than fossil fuel products.[21] The increase in the use and production of biofuels in many countries is due to the policies implemented by government.[22] Government measures put in place to support domestic biofuels production can be protectionist in nature, which can hinder the cross-border trade of biofuels. It is essential to ensure that the extent of support that is given to biofuels production has the least market distorting effects. This will ensure the establishment of a liberal global market that facilitates the international trade of biofuels. Presently, government measures supporting biofuels production has raised important considerations in international trade law.[23] Government policies and measures in Europe, USA, Canada and Brazil have been the subject of scrutiny against these countries' obligations under World Trade Organization (WTO) law.[24]

It is against this background that South Africa's biofuels industry is seeking to emerge. South Africa is a country that is highly dependent on fossil fuels for its energy needs. About 60% of liquid fuel required to meet the country's energy needs is sourced from imported crude oil.[25] About 40% is sourced from domestically produced synthetic fuels.[26] With Cabinet's endorsement in 2007 of the Biofuels Industrial Strategy,[27] govern-

19 Junginger et al. (2009: 2031).

20 Fourle (2011: 24).

21 (Ibid.).

22 Howse et al. (2006: 5).

23 (Ibid.).

24 See the following articles which have discussed government measures supporting biofuels production and their compatibility with WTO Law: Switzer & McMahon (2011); Seminerio (2007-2008); and Christian (2009).

25 Mabugu et al. (2009: 281).

26 (Ibid.).

27 Government of the Republic of South Africa (2007). The Biofuels Industrial Strategy is the first official policy document which mandates the use of biofuels in the transport sector. The Department of Energy has also recently drafted a position paper on the biofuels industry. See Government of the Republic of South Africa (2014). This position paper reinforces the policy objectives of the Biofuels Industrial Strategy. The final position paper is yet to be published by the Department of Energy. It is important to understand the policy objectives underlying the Biofuels Industrial Strategy as these policy objectives have helped to shape the measures governing the industry. The main policy objective of the Biofuels In-

ment has put in place measures to support the growth of the biofuels industry in the country. As South Africa is a member of the WTO, important considerations arise as to whether these measures are compatible with South Africa's obligations under WTO law.

B. South Africa's obligations under the WTO

The WTO has its roots in the General Agreement on Tariffs and Trade (GATT) 1947. This was a multilateral treaty which was aimed at facilitating and regulating international trade by reducing tariff barriers. South Africa acceded to the GATT 1947 in 1947.[28] Section 2 of the Geneva General Agreement on Tariffs and Trade Act 29 of 1948 approved the Executive's accession to the GATT 1947.[29] After the Uruguay Round on multilateral negotiations, the GATT 1947 was superseded by the General Agreement on Tariffs and Trade (GATT) 1994[30] and the WTO was established to regulate international trade.[31] South Africa also participated in the Uruguay Round, and became one of the first members to join the WTO.[32] South Africa had to accede to the Agreement establishing the World Trade Organization (WTO Agreement) to join the WTO. The Multilateral Trade Agreements are annexed to the WTO Agreement. All WTO members must accept the terms of the WTO Agreement and all the Multilateral Trade Agreements as a single undertaking.[33]

---

dustrial Strategy is the facilitation of rural development of former homeland areas. Other policy objectives driving this strategy include inter alia developing renewable energy resources; energy security considerations; and environmental considerations. See Government of the Republic of South Africa (2014: 13). See also Funke et al. (2009: 223).

28 Schlemmer (2004: 125).

29 Dated 18 November 1947 and published in the Government Gazette 3896 under Notice 2421. See also van Niekerk & Schulze (2000: 6) and Schlemmer (2004: 125).

30 For more information see https://www.wto.org/english/docs_e/legal_e/legalexpl gatt1947_e.htm, accessed 18 June 2014. The GATT 1994 includes the original articles of the GATT 1947 and the amendments made to this agreement as a result of the Uruguay Round of multilateral negotiations.

31 Schlemmer (2004: 125).

32 For more information on South Africa and the WTO, see http://www.wto.org/english/thewto_e/countries_e/south_africa_e.htm, accessed 14 August 2015.

33 Van den Bossche & Zdouc (2013: 109). The author makes reference to the WTO case of *Argentina – Safeguard Measures on Imports of Footwear* (20 March 1997) WT/DS22/AB/R where the Appellate Body held that the WTO Agreement

With regard to international law, South Africa follows a dualist approach.[34] The Constitution of the Republic of South Africa 1996 provides that all international agreements that are ratified or acceded to by the Executive must also be ratified by Parliament.[35] International agreements form part of the law of South Africa only when they are enacted into law by national legislation.[36] South Africa did not domesticate any of the agreements of the WTO.[37] The effect of South Africa ratifying the WTO agreements serves to bind South Africa on an international level. These agreements do not form part of domestic law.[38] The Constitution nonetheless provides that courts are obliged to take into consideration international law and international agreements even where such agreements are not part of South Africa's domestic law.[39] Schlemmer argues that:[40]

> [e]ven though the WTO agreements have not been incorporated into South African law it is clear that the government is inclined to comply with its international law obligations in terms of these agreements. This is also reflected in South Africa's actions in complying with its notification requirements in terms of the WTO.

---

and the multilateral trade agreements "represent an inseparable package of rights package or rights and disciplines".

34 Dugard (2000: 54).

35 Section 231 (2) of the Constitution.

36 Section 231 (4) of the Constitution.

37 Schlemmer (2004: 126). The author makes reference to debates of the National Assembly of 6 April 1995 (*Debates of the National Assembly (Hansard) second session – First parliament* 23 January - 11 October 1995, 642-653) and states that:
"The National Assembly adopted the report of the Portfolio Committee on Trade and Industry on 6 April 1995 – this report merely requested parliament to agree to the accession to the Marrakesh Agreement. None of the WTO agreements were ever published in the Government Gazette or incorporated into South African legislation"

38 Schlemmer (2004: 135). See also the case of *Progress Machine Offices v The South African Revenue Service and Others* 2007 4 ALL SA 1358 (SCA) para. 6 where the court commented on the status of WTO Agreements: "The WTO Agreement was approved by Parliament on 6 April 1995 and is thus binding on the Republic in international law but it has not been enacted into municipal law...No rights are therefore derived from the international agreements themselves"; See also the case of *International Trade Administration Commission v SCAW (Pvt) ltd and Others* CC 2010 ZACC 6 paras 2 and 25.

39 Section 233 of the Constitution.

40 Schlemmer (2004: 126). See also the *International Trade Administration Commission* case at para. 2. The Court states that the WTO obligations were also honoured by the enacting of legislation which governs the imposition of anti-dumping duties, countervailing and safeguard measures.

C. Statement of the problem

It is expected that with the continued rising interest in biofuels production, the international trade of biofuels will increase. South Africa is a country that is encouraging the growth of its biofuels industry. There is a need to ensure that government policies and measures governing the biofuels industry are compatible with South Africa's obligations under WTO law. Where such measures are not compatible and this affects the rights and privileges of fellow WTO members, dispute settlement proceedings may be instituted against South Africa. Countervailing, safeguarding and anti-dumping duties may also be imposed in appropriate circumstances. This study seeks to analyse the regulatory framework that governs the biofuels industry and to determine its compatibility with WTO law. Section 2 endeavours to identify the regulatory measures in place. Section 3 seeks to analyse the identified regulatory measures against WTO law. Concluding remarks and recommendations are set out in section 4.

D. The regulatory framework governing the biofuels industry in South Africa

I. Petroleum Products Act

The Petroleum Products Act 120 of 1977 (Petroleum Products Act) regulates the sale, distribution and manufacturing of petroleum products in South Africa.[41] Section 1 defines petroleum products as:

> any petroleum fuel and any lubricant, whether used or unused, and *includes any other substance which may be used for a purpose for which petroleum fuel or any lubricant may be used* (italics denote writer's emphasis)

This definition is very wide and can encompass biofuels which are used as a transport fuel. Where biofuels are manufactured for the purposes of use as a transport fuel, such manufacture and sale is governed and regulated by the Petroleum Products Act. The Minister of Minerals and Energy is empowered to fix the price at which any petroleum products may be sold or bought.[42]

---

41 Preamble of the Petroleum Products Act.

42 See generally section 2 which outlines the power of the Minister of Minerals and Energy with regards to the regulation of petroleum products in South Africa.

The Minister of Minerals and Energy is permitted, in terms of the Act, to make regulations regarding the conditions concerning the buying or selling of petroleum products made from coal, natural gas or vegetable matter.[43] The Minister is also empowered to make regulations concerning the supply, to licensed retailers, of petroleum products made from coal, natural gas or vegetable matter;[44] and can pass regulations regarding the fuel standard and specifications and blending requirements of fuel products with other substances.[45] These provisions serve as a basis for the regulation of biofuels which are produced and used as a fuel in the transport sector.

All persons who manufacture and sell petroleum products are required to be licensed by the Controller of Petroleum Products.[46] The regulations regarding petroleum products manufacturing licences[47] provide for the licensing of petroleum products manufacturers. These regulations also apply to biofuels producers.[48] A criterion for the licensing of biofuels manufacturers has also been developed by the Office of the Controller of Petroleum Products.[49] This criterion incorporates the requirements of the Biofuels Industrial Strategy and its policy objectives. The criterion also endeavours to take into account sustainability issues that surround the production of biofuels. The Office of the Controller of Petroleum Products has used this criterion to evaluate applications for biofuels manufacturing licences.[50]

43 Section 12C (1) (aA) of the Petroleum Products Act.

44 Section 12C (1) (aB).

45 Section 12C (1) (e-f).

46 Section 2A of the Petroleum Products Act. The Minister of Minerals and Energy appoints the Controller of Petroleum Products from the Public Service. The Controller has the mandate of issuing licenses in terms of the Petroleum Products Act.

47 The Regulations regarding petroleum products manufacturing licenses are Government Notice Regulation 288 of 2006 in Government Gazette 28665 (27 April 2006) as amended by Government Notice Regulation 1063 in Government Gazette 35984 (19 December 2012).

48 Gilder & Mamkeli (2014).

49 Government of the Republic of South Africa (2014: 1-2).

50 (Ibid).

II. Customs and Excise Act

The Customs and Excise Act 91 of 1964 (Customs and Excise Act) provides for the imposition of custom and excise duties. These duties serve as an important regulator in the international trade of goods. A customs duty is a monetary charge or tax on goods that are imported. It is due to be paid because goods have been imported.[51] Payment of duty grants goods market access. Excise duties are imposed on consumable products such as petroleum, tobacco and alcoholic products. Excise duties are due to be paid whether goods have been imported or produced locally.[52] The main function of excise duty is to provide revenue to government. It also aims to discourage the consumption of harmful products.[53] The WTO recognises the right of members to impose custom and excise duties on goods.

In addition to the excise duty imposed on petroleum products, the fuel levy[54] and the road accident fund levy are also imposed.[55] The obligation to pay these levies arises by virtue of the nature of the goods, whether they are manufactured locally or imported.[56] Petroleum products have been defined to include petrol, diesel and biodiesel.[57] Biodiesel is subject to a customs duty of 0.183 cents per litre. Biodiesel imports from the Southern African Development Community (SADC), European Union (EU) and European Free Trade Association (EFTA) areas do not have to pay any customs duty.[58] It is important at this stage to note that South Africa is part of the Southern African Customs Union (SACU).[59] Biodiesel is not

---

51 Van den Bossche & Zdouc (2013: 420).

52 See http://www.sars.gov.za/ClientSegments/Customs-Excise/Excise/Pages/default.aspx, accessed 23 June 2015.

53 (Ibid.).

54 The fuel levy is levied for in terms of the Schedule 1 Part 5A of the Customs and Excise Act.

55 The road accident fund levy in levied in terms of Part 5B Schedule 1 of the Customs and Excise Act.

56 Customs and Excise Act Customs and Excise Tariff Schedule Part 5A Notes para. 1 provides that "Any rate of fuel levy specified in Part 5A in respect of any goods shall apply to any such goods which are manufactured in the Republic and imported into the Republic."

57 See http://www.sars.gov.za/ClientSegments/Customs-Excise/Excise/Fuel-Petroleum-Products/Pages/default.aspx, accessed 23 June 2015.

58 Customs and Excise Act Customs and Excise Tariff Schedule 1 Part 1.

59 SACU consists of Namibia, Botswana, Lesotho, Swaziland and South Africa. These countries form a Customs Union. This means that goods can travel across the borders of these countries without being subject to pay a customs duty or excise duty.

subject to any excise duties. A customs duty of 317 cents per litre is imposed on bioethanol.[60] Bioethanol is also subject to an excise duty.[61] The Biofuels Industrial Strategy provides a fuel levy exemption of 50% to biodiesel producers.[62]

South Africa is a member of the World Customs Organization (WCO). The country has incorporated the Harmonised Commodity Description and Coding System (HS) into its domestic legislation by incorporating it into Schedule 1 of the Customs and Excise Act. The imposition of custom and excise duties is based upon the tariff classification of goods as provided for in the HS.

III. Renewable Energy Capital Subsidy Scheme

In 2005 the Renewable Energy Capital Subsidy Scheme was endorsed by the National Treasury.[63] This subsidy scheme is administered by the Department of Energy. The Renewable Energy Finance and Subsidy Office was established to assist in the administration of this subsidy scheme. The scheme provides for 16.7 cents per litre subsidy for bioethanol and 27.3 cents per litre subsidy for biodiesel up to a limit of R20 million.[64] The subsidy cannot exceed 20% of the capital cost and the minimum project size is 1 megawatt (MW) voltage.[65] However, it is not certain whether this subsidy is still available to biofuels producers.[66]

---

60 Customs and Excise Act Customs and Excise Tariff Schedule 1 Part 1.

61 Customs and Excise Act Customs and Excise Tariff Schedule 1 Part 2A.

62 Funke (2010: 70). The author notes that the fuel levy exemption would serve as an important tool to provide financial support to biodiesel producers. As the main policy objective of the Biofuels Industrial Strategy is to facilitate rural development and the creation of jobs in the energy sector, the fuel levy exemption is not available to biofuels producers who would import agricultural feedstock to produce biofuels. Government support for biofuels production is dependent on the use of agricultural feedstock produced in former homeland areas. See also Government of the Republic of South Africa (2007: 13); and Funke et al. (2009: 230).

63 Republic of South Africa (2007: 7).

64 (Ibid.).

65 (Ibid.).

66 The writer has not received any feedback or response from her numerous inquiries to the Department of Energy regarding the status of this subsidy.

IV. Comprehensive Agricultural Support Programme

In terms of the Biofuels industrial strategy, financial and capacity building support would be given to farmers growing agricultural feedstock for biofuels production under the Comprehensive Agricultural Support Programme (CASP). CASP is a support programme launched by the Ministry of Agriculture, Forestry and Fisheries in 2004.[67] The objectives of this programme are primarily to increase wealth creation in agriculture in the rural areas and to increase sustainable employment in the agricultural sector.[68] It specifically targets small-scale farmers, subsistence farmers, 'emerging farmers'[69] and beneficiaries of land reform who have received land.[70] It aims to build capacity in farmers through skills development programmes and to provide the necessary financial support.[71] Agricultural support from the production phase to marketing of agricultural goods is given, and all forms of agricultural production are able to receive support under this programme. The Ministry of Agriculture pays out grants which are administered by the various provinces. Since the implementation of CASP in 2004, a total of R750 million has been disbursed by the Ministry of Agriculture under this programme.[72] CASP grants usually take the form of once-off payments. However, projects can be structured and strategised in phases, and the allocation can be broken down accordingly and paid out over several years.[73] CASP can serve as an important vehicle which can provide financial support to farmers producing agricultural feedstock for the production of biofuels.

---

67 For further information see http://www.daff.gov.za/daffweb3/Programme/Comprehensive-Agricultural-Support-Programme, accessed 24 February 2015.

68 *Narfu Framer* (2008: 12). Other objectives of CASP include improving farming efficiency; improving household food security nationally; and improving investor confidence in the farming sector.

69 *Agriprobe* (2004). 'Emerging farmer' is the term used to describe farmers that are located in former homeland areas who were previously disadvantaged under apartheid rule.

70 *Agriprobe* (2004: 12).

71 Department of Agriculture, Forestry and Fisheries "Programmes: Comprehensive Agricultural Support Programme" *Department of Agriculture, Forestry and Fisheries.*

72 See see http://www.daff.gov.za/daffweb3/Programme/Comprehensive-Agricultural-Support-Programme, accessed 24 February 2015. In the financial years 2005-2006 and 2006-2007, R144 million and R196 million was disbursed, a total of 900 loans to the value of R24 million were approved in 2006.

73 *Agriprobe* (2004: 12-13).

V. Regulations regarding Petroleum Products Specifications and Standards

The Regulations regarding the Petroleum Products Specifications and Standards[74] prescribe the standards and specifications of petroleum products that may be sold in South Africa. Only the petroleum products which meet the specifications set out in the regulations may be sold and manufactured in South Africa.[75] Biodiesel, which conforms to the South African National Standard SANS 1935 may be used only as a transport fuel. Biodiesel may be blended with diesel from a 5% concentration to 100%.[76] Bioethanol may be blended with petrol from a range of 2% to 10% concentration. The promulgation of these regulations created legal certainty for biofuels to be blended with transport fuel in South Africa.[77]

VI. Mandatory blending regulations

To encourage the blending of biofuels with petrol and diesel, the Regulations regarding the Mandatory Blending of Biofuels with Petrol and Diesel were promulgated on 23 August 2012.[78] All petrol and diesel which is sold as transport fuel in South Africa must be blended with biofuels. Bioethanol is defined as bioethanol which meets the requirements of the South African National Standard, SANS 465. Biodiesel is defined as biodiesel which meets the requirements of the South African National Standard, SANS 1935.[79] This lays down the standard to which biodiesel and bioethanol must conform to be sold as a transport fuel. Petroleum product

---

74 Government Notice Regulation 627 in Government Gazette 28958 (23 June 2006).

75 Sections 4 and 6 (2) of the regulations regarding the petroleum products specifications and standards.

76 Section 5 of the regulations regarding the petroleum products specifications and standards.

77 Government of the Republic of South Africa (2014: 11). These regulations have been viewed as "a major step towards creating a regulatory environment that is conducive to the commercial production of biofuels."

78 Government Notice Regulation 671 in Government Gazette 35623 (23 August 2012).

79 Section 1 of the mandatory blending regulations.

manufacturers[80] are only allowed to purchase biofuels from a licensed biofuels manufacturer.[81]

The regulations set a regulated price at which petroleum product manufacturers must purchase biofuels.[82] Bioethanol can be blended with petrol from a concentration of 2% to 10%.[83] The final blend produced must conform to the South African National Standard, Unleaded Petrol, SANS 1598.[84] Biodiesel can be blended from a minimum concentration of 5 %. The final blend produced must conform to the South African National Standard, Automotive Diesel Fuel, SANS 342.[85]

All licensed biofuels manufacturers must submit records to the Controller of Petroleum Products at the end of each month of the biofuels that they have produced and sold to petroleum product manufacturers. Records must be submitted that include the amount of bioethanol or biodiesel manufactured, the amount of bioethanol or biodiesel sold to blenders, and the names of blenders to which the biodiesel or bioethanol was sold.[86] All licensed petroleum products manufacturers must also submit records to the Controller of Petroleum Products of the biofuels that they have bought from licensed biofuels manufacturers.[87] This information will assist in regulating the blending of biofuels with the country's transport fuel supply. This information will ensure that only licensed biofuels producers provide biofuels for blending. A Biofuels Implementation Committee and Blending Logistics have been established by the Department of Energy to address any problems that may arise pertaining to the mandatory blending regulations before the regulations become operational.[88] The regulations will only become effective on a date to be announced by the Minister of Energy by notice in the Government Gazette.[89] These regulations were supposed to come into operation as of 1 October 2015.[90] However, the

---

80 The term 'manufacturer' must be interpreted in terms of the Petroleum Products Act section 1 'the manufacture petroleum products also include the blending and re-refining of petroleum products'. Those who are involved in these activities are regarded as petroleum product manufacturers.

81 Mandatory blending regulations section 3.

82 Section 3 (3). The regulated price at which petroleum product manufacturers must purchase biofuels is yet to be finalised.

83 Section 3(5)(b).

84 Section 4(a).

85 Section 4(b).

86 Section 5(a)

87 Section 5(b).

88 Government of the Republic of South Africa (2014: 3).

89 Mandatory blending regulations section 6.

90 Government of the Republic of South Africa (2013).

regulations are yet to become operational as the pricing framework for biofuels is yet to be finalised.

VII. Framework for the sustainable production and use of biofuels

The sustainable production of biofuels is a topical issue. There are concerns that the production of biofuels may compete with food production and thereby cause food prices to rise. There are also concerns that biofuels production may actually contribute to global warming and impact negatively on biodiversity.[91] Various countries have responded by putting measures in place to promote the sustainable production of biofuels.[92] There are also a growing number of mandatory and voluntary sustainability frameworks and certification schemes for biofuels production which are being developed and implemented.[93] Where sustainability frameworks provide for mandatory standards which must be complied with to gain market access, this raises important considerations as to whether such frameworks are compliant and compatible with WTO disciplines.[94]

At a regional level in southern Africa, the SADC[95] has developed a regional framework for the sustainable production of biofuels in the region.[96] In December 2009, countries belonging to SADC adopted the Framework for Sustainable Biofuels.[97] This framework sets basic guidelines for the development of sustainable biofuels production in the SADC region.[98] Principles include that biofuels production should not lead to deforestation and loss of biodiversity. Biofuels production must contribute to reducing greenhouse gas emissions and promoting rural development.[99]

---

91 Cushion et al. (2013: 631-632).

92 Schut et al. (2013: 632).

93 van Dama et al. (2008).

94 For example, concerns have been raised regarding the EU's sustainability criteria for biofuels production and its compatibility with WTO law. See Ackrill & Kay (2011); Switzer S & JA McMahon (2011).

95 In southern Africa, SADC plays an important role in furthering regional trade and integration. Article 5 (1) (a) (b) of the SADC treaty provides that one of the key objectives of SADC is to "promote economic and social development ...through cooperation and economic integration." See SADC Treaty at http://www.sadc.int/files/9113/5292/9434/SADC_Treaty.pdf, accessed 13 April 2016.

96 Schut et al. (2013: 635).

97 SADC (2009).

98 Maltsoglou et al. (2013: 105).

99 SADC (2009: 3).

Many of the principles set out in the SADC framework have been taken into account by the South African government in the formation of its policy regarding biofuels production. The principles set out in the Framework for Sustainable Biofuels Production serve as general principles which governments in the SADC region are encouraged to take into account when developing policy on biofuels production in their respective countries. These principles do not form mandatory standards to which biofuels producers must adhere. It can be argued that the SADC Framework for Sustainable Biofuels Production does not raise any issues of compatibility with WTO law as these guidelines are not binding and mandatory on biofuels producers.

E. Compatibility of the regulatory measures with WTO law

Recently, there has been an increase in energy-related disputes in the WTO.[100] Most of these energy-related disputes relate to biofuels.[101] Some WTO members have also imposed countervailing and anti-dumping duties on imported subsidised biofuels.[102] This shows that the measures that countries have put in place to govern their biofuels industry are a contentious issue in the WTO. Thus it is imperative to ensure that measures implemented by South Africa to govern its biofuels industry are compliant with WTO law. It is important first to consider the international trade law rules that apply to biofuels.

---

100 Perisin (2014: 371).

101 (Ibid.: 372). There are a number of disputes in which the EU is involved before the WTO Dispute Settlement Body which relate to biofuels, including *European Union and a Member State: Certain Measures Concerning the Importation of Biodiesels*, Complainant: Argentina (17 August 2012) DS443; *European Union: Certain Measures on the Importation and Marketing of Biodiesel and Measures Supporting the Biodiesel Industry, Complainant: Argentina* (15 May 2013) DS459; *European Union: Anti-Dumping Measures on Biodiesel from Argentina, Complainant: Argentina* (19 December 2013) DS473; *European Union: Anti-Dumping Measures on Biodiesel from Indonesia, Complainant: Indonesia* (10 June 2014) DS480. See also Perisin (2014). The author attributes the large number of disputes involving biofuels to the sustainability criteria and the support measures that EU has put in place to govern the biofuels industry.

102 Harmer (2009). The author notes that "[t]he European Commission has imposed countervailing and anti-dumping duties US biodiesel imports to counter what it considers to be unfair tax incentives."

I. Biofuels under WTO law

Biofuels are considered as a renewable source of energy. However, the WTO framework does not have specific rules which address the trade in energy goods and services.[103] This is due to the fact that when the rules of the GATT were negotiated over 60 years ago, world energy demand was not a critical issue. The need specifically to address the liberalisation of trade in energy at that time was not a priority.[104] The basic rules which apply to all other trade of goods apply to the trade in energy goods. However, it has been noted that this is an undesirable situation. The energy sector is unique in various ways when compared with all other industries. This warrants the creation of special rules specifically to address this sector.[105] In relation to the international trade of biofuels, the policy issues surrounding the production and consumption of biofuels need to be addressed specifically in the context of WTO law.[106]

In assessing the application of WTO rules to biofuels, it is important to determine which WTO rules apply.[107] In the WTO framework, different rules apply to industrial goods and agricultural goods. The WTO Agreement on Agriculture (AoA)[108] provides for specific rules for goods which are considered to be agricultural goods under this agreement. Annex 1 of the AoA defines what is considered to be an agricultural good for the purposes of this agreement.

Bioethanol falls under HS Chapter 22 and is classified as an agricultural good. Bioethanol is classified with regard to its chemical composition as undenatured and denatured alcohol.[109] There is no specific sub-classification to apply to bioethanol that is used for the purposes of a fuel.[110] Biodiesel falls under Chapter 38, and is accordingly classified as an industrial good. Chapter 38 deals with "miscellaneous chemical products" and falls under section IV of the HS, which addresses "products of the chemical or allied industries".

---

103 Cottier et al. (2011: 211-212). The authors at page 211 also note that "[t]he regulation of energy international law is highly fragmented and largely incoherent."

104 Cottier et al. (2011: 212).

105 (Ibid.).

106 (Ibid.: 251).

107 Motaal (2008: 76).

108 The WTO Agreement on Agriculture is available at https://www.wto.org/english/docs_e/legal_e/14-ag.pdf, accessed 2 June 2015.

109 HS 220710 and 2207020.

110 Motaal (2008). The author notes that close inspection of the HS reveals that countries had not anticipated the growth in the trade of bioethanol as a fuel.

The different classification of biofuels products raises important considerations in the context of international trade. A consequence of this is that bioethanol and biodiesel have been placed on different legal footings.[111] Different legal regimes apply and this may hamper efforts in achieving a transparent global biofuels market.[112] Transparency is important as the biofuels industry has been subject to a wide range of governmental support measures. The different tariff classification of biofuels may impede progress in further liberalising the international trade of biofuels.[113] The Doha Ministerial Declaration advocates for "the reduction or, as appropriate, elimination of tariffs and non-tariff barriers to environmental goods and services".[114] Brazil, India and the European Union have advocated for biofuels to be classified as an environmental good.[115]

II. WTO rules on subsidies

Some of the WTO rules specifically govern the use of subsidies. Subsidies are a delicate matter in the trade relations of countries. Subsidies can be a tool used by government to implement legitimate policy objectives. Conversely, subsidies can have negative effects on the economies of other trading partners, whose domestic and export markets can be injured by unfair competition with subsidised goods.[116] WTO rules governing subsidies are laid down in Article VI and Article XVI of the GATT 1994 and the

111 (Ibid.). Mortaal notes that, "[N]ot that the rules for industrial goods are laxer than rules for the agricultural, or vice versa; but rather they are simply different. Therefore different legal regimes govern the trade policy frameworks surrounding these fuels." This is undesirable, and it does not facilitate trade liberalisation in the trade of biofuels on an equal footing. He further notes that in trade negotiations, the mandates for industrial goods and agricultural goods can be different. This is evidenced in the Doha Round of negotiations where the mandate for industrial goods and agricultural goods are different. See the WTO Hong Kong Ministerial Declaration of 2005 (WT/MIN(05)/DEC para. 24). The Declaration is available at https://www.wto.org/english/thewto_e/minist_e/min05_e/final_text_e.htm, accessed 19 March 2015.

112 Motaal (2008: 77).

113 (Ibid.).

114 Doha Ministerial Declaration of 2001 (WT/MIN(01)/DEC/1 para. 31 (iii)) The Doha Ministerial Declaration is available at https://www.wto.org/english/thewto_e/minist_e/min01_e/mindecl_e.htm#top, accessed 19 March 2015.

115 It was proposed that the definition of environmental goods should cover products such as renewable energy, including bioethanol and biodiesel.

116 Van der Bossche & Zdouc (2013: 754).

Subsidies and Countervailing Agreement (SCM Agreement),[117] which applies to both industrial and agricultural goods. The AoA however provides more specific rules for the subsidisation of agricultural goods.[118] Article 21.1 of the AoA provides that:

> [t]he provisions of the GATT 1994 and of other multilateral trade agreements in Annex 1A to the WTO Agreements shall apply subject to the provisions of this Agreement.

In *United States – Subsidies on Upland Cotton,*[119] the Appellate Body commented on the application of Article 21.1. It was held that the GATT 1994 and other multilateral trade agreements apply "to the extent that the AoA contains specific provisions dealing specifically with the same matter".[120] Hence, the SCM Agreement functions subject to the provisions of the AoA. Where the AoA contains specific provisions dealing with the same matter, the specific provisions of the AoA prevail.[121]

III. Renewable Energy Capital Scheme: The subsidy for biodiesel

The provisions of the SCM Agreement would apply to determine the legality of this subsidy as biodiesel. The Renewable Energy Capital Scheme offers a subsidy of 27.3 cents per litre for the subsidy of biodiesel production. The subsidy would qualify as financial contribution as it amounts to a direct transfer of funds.[122] The subsidy is industry-specific in that it is available to producers of a specific product. It can be argued that this subsidy, when analysed against the requirements of Article 5 of the SCM Agreement, does not have any adverse effects on the interests of other WTO members as it is a subsidy that aims to support an infant industry.

117 The Subsidies and Countervailing (SCM) Agreement is available at https://www.wto.org/english/docs_e/legal_e/24-scm.pdf, accessed 2 June 2015.

118 Van der Bossche & Zdouc (2014: 836).

119 WT/DS267/AB/R (21 March 2005).

120 *United States – Subsidies on Upland Cotton* case para. 532. The Appellate Body also made reference to the Appellate Body Report, *European Communities – Regime for the Importation, Sale and Distribution of Bananas* WT/DS27/AB/R (25 September 1997) DSR 1997:II, 591 para. 155, where the Appellate Body in the case also reached the same conclusion.

121 *United States – Subsidies on Upland Cotton* case para. 533.

122 Article 1.1 (a) SCM Agreement.

IV. The subsidy for bioethanol and CASP

The provisions of the AoA would apply to determine the legality of the subsidy for bioethanol under the Renewable Energy Capital Scheme. The Renewable Energy Capital Scheme provides for 16.7 cents per litre for the subsidy of bioethanol production. As bioethanol is classified as an agricultural product, the specific provisions of the AoA prevail over the provisions of the SCM Agreement. CASP is an important vehicle which can serve to provide financial support to emerging farmers who are involved in producing agricultural feedstock for the production of biofuels. The provisions of the AoA would also apply in determining the legality of the support offered under CASP.
The AoA provides that:[123]

> [a]ny domestic support measure in favour of agricultural producers, including any modification of such measure, and any measure that is subsequently introduced that cannot be shown to be exempt from reduction ... shall be included in the member's calculation of its current total AMS.

The level of support that a WTO member may provide to its agricultural sector may not exceed the Aggregate Measure of Support (AMS) to which that Member state has committed itself.[124] Agricultural subsidies can be classified into three different categories: green box subsidies, blue box subsidies and amber box subsidies.[125] Subsidies which are classified into the green box and blue box are subsidies which have "no, or at most minimal, trade distorting effects on production".[126] This category includes support measures for research and development. Green box subsidies and blue box subsidies are not included in a member's AMS. Where a support measure has trade-distorting effects, such a measure falls into the amber box. Such a measure must be included in a member's total AMS. It can be argued that the subsidy for bioethanol under the Renewable Energy Capital Scheme provides price support to bioethanol producers. The government must ensure that the measure of support that is provided does not exceed its total AMS. The financial support and the capacity-building sup-

123 Article 7.2 a AoA.
124 Article 6 AoA.
125 Josling & Blandford (2009: 531).
126 Annex 2, paragraph 1 of the AoA. See also Josling & Blandford (2009: 556-557). The authors at page 556 state that: "Two criteria are specified to ensure that this requirement is met: Support should be provided through a publicly funded government program and should not involve transfers from consumers; and the measures should not provide price support to producers."

port provided to farmers under the CASP programme can be justified as a green box subsidy. The financial support can be justified as "payments to producers in disadvantaged regions under region assistance programs".[127]

V. Fuel levy exemption

The Biofuels Industrial Strategy increased the measure of support given to biofuels producers by increasing the fuel levy exemption accorded to biofuels producers to 50% for biodiesel. Small-scale biodiesel producers enjoy a fuel levy exemption of 100%. The fuel levy exemption could qualify as a subsidy under Article 1.1(a)(1)(ii) of the SCM Agreement as government revenue, otherwise due, is foregone. The subsidy is industry-specific in that it is targeted at biodiesel producers.

Bioethanol producers do not pay fuel levy. One can argue that this does not qualify as subsidy in terms of the SCM Agreement. The fuel levy is levied in terms of the Customs and Excise Act, schedule 1 part 5A. It is imposed on the basis of the tariff classification of products in terms of the Customs and Excise Act which incorporates the HS. Schedule 1 part 5A provides the products which are subject to pay the fuel levy. Products which are subject to pay the fuel levy are products which are classified under tariff headings 2710.12 and 3826.00.[128] The tariff classification for ethanol, both natured and denatured, is not included in the products subject to pay fuel levy. This analysis leads to the conclusion that bioethanol is not subject to the fuel levy. There is no financial contribution in the form of government revenue due which has been foregone as there is no government revenue which would have been due in the circumstances.

However, one can argue that as bioethanol is being used as a transport fuel, bioethanol producers do have an advantage. They do not have to pay the fuel levy, whereas other products which are being used as transport fuels are subject to pay the fuel levy. The different treatment accorded to bioethanol and biodiesel in this case is a direct consequence of the different legal footing of these products under WTO law. One is categorised as an industrial good and the other is categorised as an agricultural good.

---

127 Josling & Blandford (2009: 557). See also paragraph 13, Annex 2 of the AoA.

128 Products subject to fuel levy are petrol (2710.12.02), Aviation kerosene (2710.12.07), illuminating kerosene (2710.12.15 and 2710.2.26), distillate fuel (2710.12.30), specified aliphatic hydrocarbon solvents (2710.12.37 and 2710.12.39), and biodiesel (3826.00.10 and 3826.00.90).

On the premise of the Biofuels Industrial Strategy, the support mechanism of the fuel levy exemption would only be accorded to biofuels manufacturers which are licensed in terms of the Petroleum Products Act. To be licensed as a biofuels manufacturer, biofuels producers need to comply with the requirements set out in the manufacturing license regulations and with the criteria for the licensing of biofuels producers, which have been developed by the Office of the Controller of Petroleum Products. One of the requirements is that biofuels manufacturers must source part of their feedstock from the former homeland areas. Biofuels producers who may import biofuels feedstock may not be able to benefit from the fuel levy exemption. This would be contrary to the SCM Agreement which prohibits subsidies which are dependent on the use of domestic products over imported products.[129] However, it could also be argued that the fuel exemption does not contravene the SCM Agreement as it does not create adverse effects for other WTO members.

1. The legality of fuel levy exemption under GATT 1994

Commenting on the relationship between the Multilateral Trade Agreements under the WTO, the panel in the case of *Indonesia – Certain Measures Affecting the Automobile Industry*[130] held that the Multilateral Trade Agreements were drafted with different purposes and objectives. The Agreements are complementary. Unless stated otherwise, the provisions are complementary to one another and are not mutually exclusive. This entails that a measure which may be legal under one WTO Agreement can be found to violate the provisions of another WTO Agreement.

A fundamental principle of the WTO is the National Treatment Principle. This principle stipulates that a WTO member country should not favour domestic products over like foreign products once such products have entered into their domestic market.[131] The National Treatment principle applies to both *de jure* and *de facto* discrimination.[132] Article III: 2 of the GATT provides that:

129 SCM Agreement Article 3.1 (b)
130 WT/DS54/R, WT/DS55/R, WT/DS59/R, WT/DS64/R (23 July 1998).
131 Article III:1 of the GATT 1994.
132 Van der Bossche & Zdouc (2013: 353-354).

> [t]he products of the territory of any contracting party imported into the territory of any other contracting part shall not be subject, directly or indirectly, to internal taxes of those applied to like domestic products.[133]

Where foreign imported products from a WTO member are subjected to a greater rate of tax than that applied to like domestic products in the territory of another WTO member, it can be regarded as a protectionist measure contrary to WTO principles.

In determining whether a measure is an internal tax within the scope of article III: 2, an analysis must be made of the characteristics of the measure. The obligation to pay a charge must accrue because of an internal event. The fuel levy may be regarded as an internal tax within the scope of article III: 2 as it is a levy that is due to be paid where products subject to the levy are consumed within South Africa.[134]

The Biofuels Industrial Strategy provides that, to qualify for the fuel levy exemption, a biofuels manufacturer must be licensed as a biofuels manufacturer. To be licensed as a biofuels manufacturer, a producer must make use of domestic agricultural feedstock in the production of biofuels. Use of imported agricultural feedstock in the production of biofuels is only to be allowed in exceptional circumstances. The factual implications of these requirements are that biodiesel which is imported is taxed in excess of biodiesel which is domestically produced. This results in the *de facto* discrimination of imported biodiesel.

2. Applicability of exceptions under GATT 1994 to the fuel levy exemption

Article XX provides limited exceptions in which a member may derogate from obligations under the GATT 1994. Article XX endeavours to strike an equilibrium between the competing interests of trade liberalisation on one hand and societal values and interests on the other hand.[135]

---

133 (Ibid.: 356). The Authors note that "a three tier test has been developed to determine whether there has been a violation of Article II:2: Whether the measure at issue is an internal tax/ internal charge on products; whether the imported and domestic products are like products; and whether the imported products are taxed in excess of the domestic products.

134 See http://www.sars.gov.za/ClientSegments/Customs-Excise/Excise/Fuel-Petroleum-Products/Pages/default.aspx, accessed 28 March 2016.

135 Van den Bossche & Zdouc (2013: 356).

Article XX (b) makes provision for a measure which is otherwise GATT-inconsistent where such measure is "[n]ecessary to protect human, animal or plant life or health".[136] There is a need to look at the policy objectives underpinning a measure and the design and structure of the measure to determine the applicability of article XX (b).[137]

The fuel levy exemption is an important measure in implementing the Biofuels Industrial Strategy. One of the policy objectives underlying the Biofuels Industrial Strategy has been the reduction of greenhouse gas emissions.[138] South Africa is party to the Kyoto Protocol, which is dedicated to the reduction of greenhouse gas emissions. Increased emissions of greenhouse gases can result in climate change. This can in turn have an impact on the health and life of humans, animals and plants. The fuel levy exemption encourages the use and production of biofuels, which emit less greenhouse gas. On this premise, one can argue that the fuel levy exemption qualifies as a measure necessary to protect the life and health of humans, animals and plants. However, to fully qualify as a measure justified in terms of article XX, the measure must also meet the requirements of the chapeau of article XX.[139] It is submitted that on this premise, the fuel levy would fail to be justified under article XX (b), as the fuel levy exemption is selective in its application.

Article XX (g) makes provision for a measure which is otherwise GATT-inconsistent where such measure relates to "[t]he conservation of exhaustible natural resources if such measures are made effective in conjunction with restrictions on domestic production or consumption".[140]

One of the objectives underlying the Biofuels Industrial strategy was to encourage the production and use of renewable sources of energy and to reduce the consumption of fossil fuels.[141] However, this measure would fail to be provisionally justified under article XX (g) as the fuel levy tax emption is not applied in conjunction with restrictions on domestic production or consumption of fossil fuels. From an analysis of article XX of

136 Article XX (b) of the GATT 1994. See also Van der Bossche and Zdouc (2013: 554).

137 (Ibid.).

138 Government of the Republic of South Africa (2007: 8).

139 The Chapeau of Article XX of the GATT 1994 provides for the following: "[S]uch measures should not applied in a manner which would constitute a means of arbitrary or unjustifiable discrimination between countries where the same conditions prevail, or a disguised restriction on international trade…"

140 Article XX (g) of the GATT 1994. See also Van der Bossche & Zdouc (2013: 565).

141 Government of the Republic of South Africa (2007: 8).

the GATT 1994, it can be seen that the socioeconomic policy objectives, which were the government's major driving policy objectives, cannot qualify as an exception in terms of article XX of the GATT 1994.

One could argue that the fuel levy exemption can be justified under security exceptions.[142] In international relations, national security has traditionally taken priority over trade.[143] It has been argued by some WTO members that certain industries which include the production of energy are of strategic importance.[144] One of the policy objectives underlying the Biofuels Industrial Strategy is for biofuels production to contribute to the country's energy security. One could argue that the fuel levy exemption can be justified as a security exception.[145]

VI. Mandatory blending regulations

The mandatory blending regulations provide that petroleum product manufacturers can only buy biofuels from licensed biofuels manufacturers. This means that, where biofuels are imported, they cannot be sold as blending components. This effectively shuts the market off from foreign biofuels producers who may want to export their product to South Africa. The net effect of these regulations is the violation of the National Treatment Principle. The mandatory blending regulations establish less favourable treatment for the use of biofuels that are not produced in the territory of South Africa. This is inconsistent with the provisions of article III: 4 of the GATT 1994. These regulations also establish a local content requirement which is inconsistent with provisions of article III: 5 of the GATT 1994.

The blending regulations have been promulgated to create the demand for biofuels which will in turn encourage biofuels production. Government policy objectives of developing energy security could justify the inconsistency of this measure with the National Treatment Principle.

---

142 The GATT 1994, Article XXI (b) provides that nothing in the GATT 1994 shall be construed as:
"[t]o prevent any contracting party from taking any action which it considers necessary for the protection of its essential security interests."

143 Van der Bossche & Zdouc (2013: 595). The authors state that "[s]tates may consider it necessary to restrict imports in order to protect strategic domestic production from import competition." The decision as to which domestic production qualifies as strategically important varies from different countries.

144 Van der Bossche & Zdouc (2013: 595-596).

VII. Technical standards governing the production of biofuels

Biofuels standards and specifications standards lay down product characteristics and their related process and production methods for biofuels production.[146] WTO law recognises the right of its members to put in place technical regulations and standards to ensure various social and economic goals. However, technical standards must not create unnecessary barriers to international trade.[147] The regulations regarding the petroleum products specifications and standards and the mandatory blending regulations are technical standards which govern biofuels production in South Africa. They prescribe the chemical and the physical properties that biofuels should have if they are to be sold in South Africa. The chemical and physical properties used to define and describe biofuels are referred to as specifications.[148]

Biofuels standards can have a major effect on the international trade of biofuels. There is no internationally recognised standard for bioethanol or biodiesel. Countries have biofuels standards which may set different specification requirements. This can hinder international trade, as biofuels which do not conform to the specification requirements may not be permitted to be sold in a country.[149] In 2007, Brazil, USA and the EU established a Tripartite Task Force comprising experts from standards developing organisations to review the different standards and specifications in place for biodiesel and bioethanol.[150] The Tripartite Task Force produced a White Paper on Internationally Compatible Biofuels Standards.[151] The ultimate goal of the taskforce was to work towards establishing compatible biofuels standards which would not hinder the trade of biofuels.[152] The taskforce identified 16 specifications for bioethanol and 24 specifications

146 Oosterveer & Mol (2009: 73).

147 The Preamble of the Techincal Barriers to Trade Agreement states that:
"[N]o country should be prevented from taking measures necessary to ensure the quality of its exports, or for the protection of human, animal or plant life or health, of the environment, or for the prevention of deceptive practice, at the levels it considers appropriate, subject to the requirement that they are applied in a manner which would constitute a means of arbitrary or unjustifiable discrimination between countries where the same conditions prevail or a disguised restriction on international trade."

148 Junginger et al. (2009: 2032).

149 Swinbank (2009: para. 6.3).

150 Tripartite Task Force (2007).

151 (Ibid.).

152 Para. 1.1.

for biodiesel. The specifications were ranked in three categories: specification requirements which are similar in the different countries; specification requirements which can be aligned in a short-term period of one year; and specification requirements which are fundamentally different.[153] The findings of the taskforce revealed that the specification standards for bioethanol were generally compatible.[154] It was concluded that different bioethanol standards and specifications do not constitute a technical barrier to the international trade of bioethanol.[155] However, only six of the 24 specifications for biodiesel were found to be compatible.[156] This is due to the fact that the different agricultural feedstock from which biodiesel may be made may cause differences in the chemical composition of biodiesel.[157] The report noted that biodiesel standards could form a technical barrier to trade.[158]

As noted above, biodiesel used as transport fuel in South Africa has to conform to the South African National Standard, SANS 1935. Bioethanol sold as transport fuel in South Africa must conform to the South African National Standard, SANS 465. Though these standards are voluntary,[159] they are made a compulsory requirement as legislation has incorporated them as a benchmark standard for biodiesel and bioethanol. SANS 1935 has been criticised for having a narrow definition of biodiesel, as it only caters for biodiesel made from vegetable oils, and excludes biodiesel made from other oils.[160]

These standards may be regarded as technical regulations as defined in terms of Annex 1 of the Technical Barriers to Trade Agreement (TBT Agreement).[161] Article 2.2 of the TBT Agreement provides that members

153 Para. 1.1.

154 Paras 1.23 and 1.2.3.1.

155 Para. 1.2.3.1.

156 Para. 1.2.2.1.

157 Junginger et al. (2009: 2032).

158 Tripartite Task Force (2007).

159 Nolte (2007: para. 2.4.2.).

160 (Ibid.). The author also notes that SANS 1935 only

"[s]pecifies the properties of biodiesel to be used directly as a pure fuel without blending...It does not take into account the effects of dilution in blends, and requires that the same requirements to be applied to the biodiesel that is meant for blending."

161 The Technical Barriers to Trade Agreement is available at https://www.wto.org/english/docs_e/legal_e/17-tbt.pdf, accessed 2 June 2015. In the *European Communities – Trade Description of Sardines* WT/DS231/AB/R (23 October 2002), the Appellate Body laid down a test for establishing whether a measure is a technical regulation under the TBT Agreement at para. 176:

must endeavour to make sure that technical regulations which are implemented do not have the effect of creating unnecessary obstacles to trade. Members are allowed to implement technical regulations where such measures are implemented to fulfil legitimate policy objectives. The policy objectives set out in article 2.2 are not exhaustive.[162]

The need for biofuels specification standards can be considered a legitimate objective. Consumers have no way of ascertaining the quality of these fuels by mere examination.[163] If biofuels used are not of the prescribed quality, they could cause damage to vehicles.[164] Hence, biofuels specifications and standards are necessary to protect the consumer. However, it must be recognised that these standards and specifications can become technical barriers to the trade of biofuels. There is a need for the South African government to carry out a survey of the different biofuels specification standards that govern biofuels internationally. There is also a need to ensure that the biofuels specifications and standards that are applied in the local biofuels industry are compatible with international standards that are being developed. South Africa must take an active part in discussions on establishing internationally recognised biofuels specifications and standards.

F. Conclusion and recommendations

A critical analysis of the measures regulating the biofuels industry has revealed that most of the measures are WTO-consistent. Measures which can be classified as subsidies can be argued to be consistent with WTO law. They do not create any adverse effects in the domestic industry of other WTO members. These measures can be justified on the premise that the biofuels industry is still an infant industry. Support measures given to agricultural feedstock production for the production of biofuels however raises important considerations. The South African government must ensure that the support given under these measures does not exceed its AMS.

"The measure must apply to an identifiable group of products;
The measure must lay down product characteristics;
Compliance with these products characteristics must be mandatory."

162 Van den Bossche & Zdouc (2013: 875). The authors maintain that other policy objectives can be considered to be legitimate and within the scope of article 2.2.

163 Kerr & Loppacher (2005: 57).

164 (Ibid.: 58).

The fuel levy exemption and mandatory blending regulations raise important concerns of compatibility with GATT 1994 obligations. Both these measures violate the National Treatment Principle. However, an argument can be made that such measures can be justified under article XXI of the GATT 1994. The following reforms to international trade law and South Africa's national law are proffered:

I. Reform in the national context

1. Reform of the fuel levy exemption

The fuel levy exemption for biodiesel violates the National Treatment Principle set out in Article III of the GATT 1994. Essentially, a biofuels producer can only benefit from the fuel levy exemption when agricultural feedstock is sourced from designated former homeland areas.

While this measure aims to uphold the government's policy objectives of rural development, its application results in *de facto* discrimination. Where a foreign biofuels producer decides to export biodiesel to South Africa, it will be subject to a higher fuel levy rate, as the producer cannot qualify for the fuel levy exemption. It is recommended that the requirements to benefit from the fuel levy exemption be changed. The fuel levy exemption should be extended to all biofuels producers who produce biodiesel. It should not be confined only to biodiesel producers who have sourced their agricultural feedstock from the former homeland areas.

2. Reform to the mandatory blending regulations

The mandatory blending regulations also violate the National Treatment Principle. These regulations only allow for the blending of petrol and diesel with biofuels that have been locally produced. Petroleum product manufacturers may only purchase biofuels from licensed biofuels manufacturers. Biofuels which are imported cannot be sold and used as a blending component. This effectively closes off the biofuels market from foreign competition. This measure is protectionist in nature and results in *de facto* discrimination against foreign biofuels producers who may want to export their biofuels to South Africa. It is recommended that the mandatory blending regulations be amended to allow for biofuels which are imported to be used as a blending component where they conform to the quality standards set out in the regulations.

II. Reform in international trade law

Biofuels are a renewable source of energy. However, biofuels are not categorised by the WTO on the basis of their use as a fuel. The various biofuels are classified differently by the WTO on the basis of their chemical composition. Biodiesel and bioethanol are placed on a different legal footing under the WTO. Industrial goods are subject to stricter rules than agricultural goods for the purposes of subsidisation. This is an undesirable situation. It is recommended that changes be made to the HS so that it can properly accommodate biofuels.[165] Biofuels should be classified on the basis of their use as a fuel. While the amendment of the HS would be a welcome step, amending the HS is a lengthy process which may take more than seven years.[166]

WTO law does not have any specific rules which address the trade in energy goods and services. The production and trade of energy goods is governed generally by the GATT 1994. It has been recommended that future negotiations in the WTO should work towards establishing a comprehensive agreement on energy.[167] Negotiations will facilitate the proper classification of energy goods. In this context, biofuels will be properly classified. It is also suggested that the future agreement on energy should adequately deal with the issue of subsidies for energy products. The present framework of WTO law does not make adequate provision for the subsidisation of renewable energy products.[168] The number of subsidies being made in respect of the renewable energy sector globally is growing as a policy response to energy security concerns and climate change.[169]

---

165 (Ibid.: 57).

166 (Ibid.). The authors note that the process to amend the HS is extremely complex. The process begins with private firms and industries which lobby their government to amend the HS. The government then presents a proposal to the WCO's Review Committee. The proposal goes through various review processes before an agreement is reached and an amendment is drafted. Countries have six months in which to raise any objections to the proposed amendment. The implementation of the amendment is a two and a half year process. This time period of implementation is given to align state practice with the new law.

167 Cottier et al. (2011: 221).

168 (Ibid.: 225).

169 (Ibid.: 226).

WTO law should be seen to evolve and respond to these challenges. Adequate policy space for countries with the framework of WTO law must be made in this regard.

There is a need to develop internationally recognised biofuels product specifications and standards. It is possible to produce biofuels in many different ways. The different inputs or different production methods can create biofuels which have different specifications.[170] As the biofuels industry continues to grow and develop, governments are likely to be subjected to a lot of pressure to implement measures to protect local biofuels markets.[171] The different biofuels specifications and standards can be taken advantage of and be used by governments as a protectionist measure to block imports.[172] There is a need to urgently address this to facilitate the international trade of biofuels.

Bibliography

Ackrill, R & A Kay, 2011, EU biofuels sustainability standards and certification systems – how to seek WTO compatibility, *Journal of Agricultural Economics* 62(3), 551.

Agriprobe, 2004, CASP: further support for land reform, *Agriprobe* 1(3), 12, at http://reference.sabinet.co.za/webx/access/electronic_journals/agriprob/agriprob_sep_2004_a8.pdf, accessed 24 February 2015.

Chakauya, E, G Beyene & RK Chikwamba, 2009, Food production needs fuel too: perspectives on the impact of biofuels in southern Africa, *South African Journal of Science* 105, 174.

Christian, K, 2009, Worth keeping around? The United States biofuels policies and compliance with the World Trade Organisation, *Georgia Journal of International and Comparative Law* 38, 165.

Cottier, T, G Malumfashi, S Matteotti-Berkutova, O Nartova, J De Sepibus & Z Bigdel, 2011, "Energy in WTO law and policy", in: Cottier, T & P Delimatsis (eds), *The prospects of international trade regulation: from fragmentation to coherence*, Cambridge, Cambridge University Press, 211.

Cushion, E, A Whiteman & G Dieterle, 2009, *Bioenergy development: issues and impacts for poverty and natural resource management*, Washington, The World Bank, at http://siteresources.worldbank.org/INTARD/Resources/Bioenergy.pdf, accessed 20 October 2016.

Dugard, J, 2000, *International law: a South African perspective*, Cape Town, Juta.

170 Kerr & Loppacher (2005: 59).

171 (Ibid.).

172 (Ibid.: 58). The authors note that:
"[D]iffering standards can restrict or even completely block international suppliers from competing with domestic producers, products standards often become a tool of groups that seek protection."

Fourle, H, 2011, Adding fuel to the fire: a macro perspective on biofuels production in South Africa, *Agriprobe* 8, 22.

Funke, T, 2011, *Biofuels production in South Africa: the games, the cost of production and policy options*, Thesis (DCom), University of Pretoria at http://www.repository.up.ac.za/bitstream/handle/2263/24639/thesis.pdf?sequence=1&isAllowed=y, accessed 20 October 2016.

Funke, T, PG Strauss & F Meyer, 2009, Modeling the impacts of the industrial biofuels strategy on South African agricultural and biofuels subsectors, *Agrekon* 48(3), 223.

Gilder, A & M Mamkeli, 2014, Biofuels in South Africa, *ENSafrica*, 5 March 2014, at https://www.ensafrica.com/news/biofuels-in-South-Africa?Id=1339, accessed 15 June 2016.

Government of the Republic of South Africa, Department of Minerals and Energy, 2007, Biofuels industrial strategy of the Republic of South Africa, at http://www.energy.gov.za/files/esources/renewables/biofuels_indus_strat.pdf(2).pdf, accessed 30 April 2016.

Government of the Republic of South Africa, Department of Energy, 2013, Media statement: Mandatory blending of biofuels with petrol and diesel to be effective from 1 October 2015, at http://www.energy.gov.za/files/media/pr/2013/MediaStatement-Mandatory-Blending-of-Biofuels-with-Petrol-and-Diesel.pdf, accessed 13 June 2015.

Government of the Republic of South Africa, Department of Energy, 2014, Draft position paper on the South African biofuels regulatory framework, at http://www.gov.za/sites/www.gov.za/files/37232gen24.pdf, accessed 4 March 2016.

Harmer, T, 2009, *Biofuels subsidies and the law of the WTO*, International Centre for Trade and Sustainable Development Issue Paper No. 20, available at http://www.ictsd.org/downloads/2012/02/biofuels-subsidies-and-the-law-of-the-wto.pdf, accessed 20 October 2016.

Howse, R, P van Bork & C Hebebrand, 2006, *WTO disciplines and biofuels: opportunities and constraints in the creation of a global marketplace*, IPC Discussion Paper, at http://www.agritrade.org/Publications/DiscussionPapers/WTO_Disciplines_Biofuels.pdf, accessed 10 March 2016.

Josling, T & D Blandford, 2009, "Biofuels subsidies and the green box", in: Meléndez-Ortiz, R, C Bellmann & J Hepburn (eds), *Agricultural subsidies in the wto green box: ensuring coherence with sustainable development goals*, Cambridge, Cambridge University Press, 531.

Junginger, M, J Dame, S Zarrill, F Ali Mohamed, D Marchal & A Faaij, 2009, Opportunities and barriers for international bioenergy trade, *Energy Policy* 39, 2028.

Kerr, WA & LJ Loppacher, 2005, Trading in biofuels – will International trade law be a constraint?, *Current Agriculture, Food and Resource Issues* Number 6, at http://ageconsearch.umn.edu/bitstream/46024/2/kerr6-1%5b1%5d.pdf, accessed 20 October 2016.

Mabugu, R, M Chitiga & H Amusa, 2009, The economic consequences of a fuel levy reform in South Africa, *South African Journal of Economic and Management Sciences* 12(3), 280.

Maltsoglou, I, T Koizumi & E Felix, 2013, The status of bioenergy development in developing countries, *Global Food Security* 2(2), 104.

Motaal, DA, 2008, The biofuels landscape: Is there a role for the WTO?, *Journal of World Trade* 42(1), 61.

Narfu Framer, 2008, Understanding CASP: focus on finance, 12 *Narfu Farmer* 25, at http://reference.sabinet.co.za.ez.sun.ac.za/webx/access/electronic_journals/ac_nafu/ac_nafu_n12_a11.pdf, accessed 24 February 2015.

Nolte, M, 2007, *Commercial biodiesel production in South Africa: a preliminary economic feasibility study*, Master of Science in Chemical Engineering Thesis, Stellenbosch University, at http://scholar.sun.ac.za/handle/10019.1/1797, accessed 20 October 2016.

Oosterveer, P & APJ Mol, 2009, Biofuels, trade and sustainability: a review of perspectives for developing countries. Biofuels, bio-products and bio-refining, *Wiley InterScience*, 66 at http://onlinelibrary.wiley.com/doi/10.1002/bbb.194/epdf, accessed 15 April 2016.

Perisin, T, 2014, Pending EU disputes in the WTO: challenges to EU energy law and policy, *Croatian Yearbook of European Law and Policy* 10, 371.

Rosillo-Calle, F & FX Johnson, 2010, "Introduction", in: Rosillo-Calle, F & FX Johnson (eds) *Food versus Fuel: an informed introduction to biofuels*, London, Zed Books, 1.

SADC / Southern African Development Community, 2009, Framework for sustainable biofuels, available at http://www.energy.gov.za/files/policies/petroleum/Annexure-D-SADC-Framework-for-Sustainable-Biofuels-Dec-2009.pdf, accessed 13 April 2016.

Schlemmer, EC, 2004, South Africa and the WTO ten years into democracy, *South African Yearbook of International Law* 42(1), 125.

Schut, M, N CunhaSoares, G van de Ven & M Slingerland, 2013, Multi-actor governance of sustainable biofuels in developing countries: the case of Mozambique, *Energy Policy*, 631.

Seminerio, FA, 2007-2008, A tale of two subsidies: how federal support programs for ethanol and biodiesel can be created in order to circumvent fair trade challenges under World Trade Organisation rulings, *Penn State International Law Review* 26(4), 963.

Swinbank, A, 2009, *EU support policies for biofuels and bioenergy, environmental sustainability criteria and trade policy*, ICTSD Global Platform on Climate Change, Trade Policies and Sustainable Energy, Issue Paper No. 17.

Switzer, S, 2012, "Biofuels, food security and the WTO Agreement on Agriculture" in: McMahon, JA & MG Desta (eds) *Research handbook on the WTO Agriculture Agreement: new and emerging issues in international agricultural trade law*, Cheltenham, Edward Elgar, 250.

Switzer, S & JA McMahon, 2011, EU Biofuels Policy – Raising The Question of WTO Compatibility, *International and Comparative Law Quarterly* 60, 713.

Tripartite Task Force, 2007, *White paper on internationally compatible biofuel standards tripartite task force: Brazil, European Union and United States of America, at* /http://ec.europa.eu/energy/res/biofuels_standards/doc/white_paper_icbs_final.pdfS, accessed 1 April 2016.

Tripartite Task Force: Brazil, European Union and United States of America, 2007, *White Paper on Internationally Compatible Biofuel Standards*, at https://www.nist.gov/sites/default/files/documents/public_affairs/releases/biofuels_report.pdf, accessed 20 October 2016.

van Dama, J, M Jungingera, A Faaija, I Jürgensb, G Bestb & U Fritschec, 2008, Overview of recent developments in sustainable biomass certification, *Biomass and Bioenergy*, 749 32(8), at http://www.unece.lsu.edu/certificate_eccos/documents/2008/ce08_35.pdf, accessed 20 October 2016.

Van den Bossche, P & W Zdouc, 2013, *The law and policy of the World Trade Organization* (3rd edition), Cambridge, Cambridge University Press.

van Niekerk, JP & WG Schulze, 2000, *The South African law of international trade: selected topics*, Pretoria, SAGA Legal Publications.

PART IV:
BIOFUELS POLICY IN ZIMBABWE

Biofuels policy in Zimbabwe – an overview

Eddie G Cross, Dip Agric (Gwebi), Bsc Econ (London), Member of Parliament for Bulawayo South Constituency, Zimbabwe

A. Background

In 1966 the Government of Rhodesia was placed under mandatory UN sanctions by the UN Security Council. This decision followed the move by the Government of Rhodesia to unilaterally declare independence from the United Kingdom. In support of the sanctions programme the United Kingdom mounted a naval blockade of the Port of Beira in Mozambique in an effort to prevent the delivery of crude oil to the terminal at Beira for onwards delivery by pipeline to the oil refinery at Feruka in the Eastern Highlands of Rhodesia.

The blockade was effective and the country was forced to rely on South Africa for its liquid fuels requirements. This was costly and in addition, provided the South African Government with greater leverage over the Rhodesian regime. The Government then ordered the Triangle Sugar Company in the South East Lowveld, to look into the production of ethanol from sugar byproducts. This was done and within 18 months, the Company began to produce ethanol for blending at 25% inclusion with petrol imported from South Africa.

This programme was continued until 1980 when the country came to Independence and when sanctions were lifted. The Beira/Mutare pipeline was repaired and renovated and was able to resume the delivery – this time not of crude oil, but of refined products to the now defunct oil refinery for distribution.

Because the cost of ethanol was much higher than the landed cost of fuel, the Triangle Sugar Company switched production from ethanol to alcohol for the distilling industry and for export. The country resumed the use of pure petrol manufactured from crude oil and sold on world markets. This reduced the demand on South African resources at a time when that country was itself under political pressure and was beginning to feel the influence of rising demand on their own refineries and oil from coal facilities.

Despite shortages of foreign exchange for imports and growing demand, little was done about biofuels for two decades after Independence in 1980. Then, after 2000, the accelerating collapse of the formal economy in Zimbabwe associated with persistent fuel shortages began to stimulate interest in the biofuels sector. The first project to receive attention was a proposal to grow the oilseed plant Detropha which thrives in semi arid regions and is potentially a useful cash crop for the rural peasant population.

The Industrial Development Corporation was instructed to build a plant to process the oil into diesel and this was undertaken with technical and financial assistance from the Far East at a cost of about $14 million. The government undertook a national programme to encourage the production of Detropha and initially the results were encouraging. However the seed is highly poisonous and the shrub unusable for any other purpose and significant production volumes were never achieved. This may have been partly due to the radical land reform programme launched in 2000 which forced all existing commercial farmers off the land and reallocated it on a patronage basis to Party aficionados.

Whatever the reasons, the plant remains closed and unused on the outskirts of Harare and little or no interest in the shrub remains.

The second initiative was the brainchild of a local business person who had made a large fortune in the motor and transport industry and proposed that he lease certain State owned estates along the major rivers in the South East of the country for sugar cane production specifically as a raw material for ethanol production.

In the early part of the 21st Century the Company he formed for this purpose – Green Fuels (Private) Limited, entered into a 20 years BOT lease over two Estates along the Save River – Middle Save and Chisumbanje. Within 3 years these Estates had been renovated and some 25,000 hectares of sugarcane planted. Within 5 years an ethanol plant, capable of supplying 5 million litres of high quality ethanol per month to the local market had been constructed with all ancillary features including an electricity generating power plant using local byproducts and the required storage facilities.

The total cost of this initiative was put at $600 million which was probably an inflated figure, but never the less the results were astonishing – yields of cane reached 150 tonnes per hectare and the most modern farming and harvesting techniques were employed. The initiative also came at a time of rising crude oil prices which reached $120 a barrel just as the project reached its planned output levels.

Encouraged by these market movements, the company announced ambitious plans for produce enough ethanol to provide southern African States with ethanol to blend with petrol to give consumers a choice of blended products ranging from 5% inclusion to 85% inclusion. The plan was to harness water from 5 major rivers in the South East and to construct a number of major new dams to hold back flood waters for irrigation purposes. Eventually the company planned to bring 150,000 hectares under production.

This investment was in many ways stimulated by the success of the Brazilian policies designed to give sugar cane farmers the choice of sugar or ethanol for use in the domestic markets for petrol. As in Brazil it was contemplated that a whole industry might develop for motor vehicles specially adapted to ethanol as fuel.

A major motivation given for the project was that this initiative was eco friendly and would reduce internal combustion engine emissions that were having a negative impact on climate change. Secondary considerations were reducing the demand for foreign exchange to buy petroleum products abroad and postponing major investments in regional petroleum refinery capacity or new oil from coal or gas capacity. The initiative was expected to create tens of thousands of jobs in the rural economy with many more in associated and secondary industries.

Since 2012 the project has run into severe headwinds and these can be summarized as follows:

B. The collapse of global crude oil prices

Since the super cycle in commodity prices came to an end in 2012, the price of crude oil has slumped from a high of $115 to $120 a barrel to $35 to $45 a barrel. This has led to a consequential reduction in refined fuel prices accentuated by a global surplus in refinery capacity. The Brazilian ethanol industry has struggled to cope with this development and output has plummeted.

In Zimbabwe, the single ethanol producer, Green Fuels (Pvt) Ltd. has been unable to reduce its prices and has struggled to maintain viability even at heavily subsidized price levels. The value of the subsidy to the industry is estimated at $15 to $20 million a month or some $0.20 cents a litre for petrol. This is paid for by maintaining a price for petrol in local markets at much higher levels than in other countries in the region. Despite this the company is known to be struggling financially.

C. The impacts of climate change

Zimbabwe is one of those countries that is expected to suffer most from impending climate changes as a consequence of rising temperatures. Projections suggest that the South East of the Country will become much drier overall while the North and West and the Eastern Highlands become wetter overall. This associated with greater variability in local weather patterns.

The area associated with the ethanol project is spread across the region that is likely to be most affected. Sugar cane production in this hot, dry region takes a great deal of water and it is estimated that each litre of ethanol uses 300 to 400 litres of water in production. This could be reduced by changes to the irrigation systems used but this would further increase the cost of production and make it even more difficult to market on a sustainable basis.

It would be very much better for this irrigation potential to be used for alternatives such as maize. The market margins on maize and other similar crops are much higher than that for sugar cane and ethanol and would have a greater impact on basic staple food supplies and potential exports. It would also involve a much more efficient and economical use of available water. The problem for both sugar and ethanol producers is that the industrial plant investment associated with sugar cane production is so great that they can ill afford to switch crops.

D. The marginal impacts on global warming

At 15 to 25% inclusion in petrol, the impact of the use of ethanol in the motor industry is marginal at best. In fact the generation of CO_2 from burning sugar cane prior to reaping (not practiced by Green Fuels) is much greater and when coupled to wild fires in Africa, pales into insignificance. It could be argued that the cane itself is a carbon sink but this needs a great deal more research to justify this form of cropping. Forestry in all its forms might be more effective.

E. The lower efficiencies of combustion

There is little doubt that ethanol diluted petrol is very much less efficient as a fuel and that consumption rises when it is employed. If this factor is taken into account the price of ethanol should be much lower than the price of petrol to justify inclusion. Taken together with the cost of production and the relatively low impact on the production of harmful gasses, this makes the production of ethanol very difficult to justify.

F. The alternative feed stocks

In the United States the major feed stock for biofuels production is maize or corn. The average price of this commodity in the US is about half of the cost of production in Africa. The main reasons for this are the use of genetically modified crop varieties and the use of advanced production technology and practices. For example the top yield achieved last year in the main corn belt in the USA was 34 tonnes per hectare. This compares with average yields of 300 kilograms in Zimbabwe and top yields of 19 tonnes per hectare using hybrid varieties.

A hectare of irrigated maize using good management practices would cost about $1,800 per hectare to produce in Zimbabwe. At 19 tonnes a hectare yield this is approaching the potential price for maize in Zimbabwe about that of the price of corn in the States of about $170 per tonne. At these prices the production of ethanol would be potentially possible at prices which would enable entry to the local market. However with maize as the basic staple diet and presently priced at $300 a tonne or more, it is unlikely that this situation can be brought into play in the biofuels game.

G. Conclusion

From this overview of biofuels experience in Zimbabwe since the year 2000, it is very difficult to make a case for the industry. Should crude oil prices return to $120 to $140 a barrel, this might change, but even at these prices an economic justification would be difficult to make. It would then be the case for the reduction in harmful gasses that would have to be employed and take precedence.

Developing the biofuels sector in Zimbabwe

Tjaša Bole-Rentel (MSc), World Wide Fund for Nature (WWF), South Africa

A. Introduction

I. Macroeconomic and policy context of biofuel development in Zimbabwe

After nearly two decades of contraction, the Zimbabwean economy finally returned to growth in 2009, following stabilisation of the political situation and the introduction of the multicurrency system, which ended hyperinflation and returned a certain degree of liquidity to the economy. In the period 2009–12, the Zimbabwean economy recorded high average growth rates of around 10%.[1] Since 2012, however, the economy has experienced another slowdown due to deteriorating terms of trade, adverse weather and renewed policy instability. Economic growth rate declined to 4.5% in 2013 and 3.2% in 2014, and is projected to be only around 2–3% in 2015 and 2016.[2] Exports contracted and are expected to remain depressed in the short-term, owing to a stronger US dollar and other factors affecting the competitiveness of Zimbabwean businesses. The negative trade balance is the main reason for the high current account deficit at approximately 23% of GDP (in 2014)[3] and is not likely to improve materially in the short term, unless there is substantial improvement in the performance of the domestic manufacturing sector.[4] The existing deficit is largely financed by short-term capital inflows and remittances since reserves are inadequate

1 See http://www.worldbank.org/en/country/zimbabwe/overview, accessed 16 May 2015; and http://www.africaneconomicoutlook.org/en/country-notes/southern-africa/zimbabwe/, accessed 16 May 2015.

2 See http://www.africaneconomicoutlook.org/en/country-notes/southern-africa/zimbabwe/, accessed 16 May 2015; and http://www.worldbank.org/en/country/zimbabwe/overview, accessed 16 May 2015.

3 See http://www.africaneconomicoutlook.org/en/country-notes/southern-africa/zimbabwe/, accessed 16 May 2015.

4 African Development Bank (2013).

and foreign direct investment (FDI) levels remain persistently low.[5] Inflation has turned to deflation, which now stands at around -1%, reflecting continued weakening of the South African rand and depressed aggregate demand.[6] Worryingly, industrial capacity utilisation continues to decline, and is currently estimated at only about one-third of installed capacity, owing to underproduction and lack of competitiveness.[7] As a result of low activity in the formal sectors, around 80% of the workforce is engaged in informal economic activities.[8]

The agriculture sector is still the backbone of the country's economy, accounting for 12% of GDP in 2013.[9] In 2014, it posted a robust growth of 23%, but was expected to slow down to some 3.4% in 2015 as a result of inadequate financing and flooding in some areas.[10] Agricultural products still represent about a quarter of Zimbabwean exports.[11]

Poverty levels in the country remain high, with over 70% of the population living in poverty and over 20% living in extreme poverty.[12] Combined with the effect of recurrent droughts and occasional floods, this results in food insecurity among a high number of households across the country.

While electricity supply continues to be constrained, the liberalisation of the fuel market in 2008 allowed for private companies to import fuel into Zimbabwe, which now means that fuel is widely available. Despite the international oil price being low at present, the cost of fuel at the retail level remains high owing to a number of duties and levies on fuel sales, a crucial source of revenue for the highly constrained national fiscus.

In the macroeconomic context described above, biofuels represent an opportunity to address some of the major challenges facing the Zimbabwean economy, namely:

---

5 See http://www.worldbank.org/en/country/zimbabwe/overview, accessed 16 May 2015.

6 (Ibid.).

7 See http://www.africaneconomicoutlook.org/en/country-notes/southern-africa/zimbabwe/, accessed 16 May 2015; and http://www.afdb.org/en/countries/southern-africa/zimbabwe/zimbabwe-economic-outlook/, accessed 16 May 2015.

8 African Development Bank (2013).

9 See http://www.worldbank.org/en/country/zimbabwe/overview, accessed 16 May 2015.

10 (Ibid.).

11 Monyau & Bandara (2015).

12 See http://www.worldbank.org/en/country/zimbabwe/overview, accessed 16 May 2015; and African Development Bank (2013).

- Savings of foreign currency and thus improvement of the country's balance of payments;
- Creation of a large market for agricultural products, which supports further expansion of the agricultural sector and creation of economic opportunity in the rural areas; and
- Enhancing energy security, especially in the transport sector.

In addition, domestically produced biofuels could:

- Contribute to climate change mitigation through reducing greenhouse gas emissions; and
- Improve air quality by reducing other vehicular emissions.

To achieve the benefits named above, a clear, long-term national biofuels policy needs to be put in place and lack thereof has been identified as one of the main factors preventing the development of a domestic biofuel industry in Zimbabwe.[13] In addition, it is important to recognise that biofuels can have potential negative effects across the value chain from production to marketing, if there are no clear guidelines at policy level. To meet the twin challenge of contributing to the country's sustainable energy supply without threatening its environmental integrity or food supply, the process of developing a national biofuel policy for Zimbabwe started in earnest in 2015.

II. The biofuels story so far

The production and use of biofuels in Zimbabwe has been going on for decades. Already in the 1970s, the country has been producing biodiesel based on jatropha, a plant which was first introduced in the 1940s for its exploitation for soap and lubricant production. The usage of jatropha as a biofuel feedstock declined in the 1980s.

At the same time (1980s), Zimbabwe started producing bioethanol using molasses as feedstock at the Triangle Ethanol Plant, which was the first production facility of its kind in Africa. The bioethanol plant produced close to 120,000 litres per day and 40 million litres per year. Petroleum blending was done at a rate of 12–15% with ethanol. The production plant ceased operation in 1992 owing to severe drought conditions which

13 Moyo et al. (2014).

led to failure to access the feedstock, while at the same time unblended fuel also became cheaper.[14]

In 2005, the country once again relooked at the prospect of using the jatropha plant for biodiesel production. The national biodiesel feedstock production programme was launched with the aim of establishing jatropha plantations right across the country to ensure sustained supply of the feedstock. The National Oil Company of Zimbabwe (NOCZIM) was mandated to spearhead the project. NOCZIM distributed free seeds to farmers in areas such as Mutoko and entered into a purchase contract with them for the offtake of oilseed, but, owing to financial constraints, NOCZIM cancelled the contracts unilaterally and left the farmers frustrated.

Soon after, Finealt Engineering was established as the lead agency in biodiesel development by government. Finealt built another biodiesel processing plant in Mutoko with a capacity of about 10,000 litres a day. The plant was based on sourcing most of the feedstock from the smallholder farmers in the area. In addition, Finealt had entered into a joint venture with an Italian corporation to cultivate 10,000 hectares of jatropha. On the basis of the perceived and promised benefits of a lucrative market for biodiesel feedstock, many people and institutions started growing *Jatropha curcas* in a rather uncoordinated fashion. The term 'Jatropha – broom of poverty' was even coined to reflect people's perception of how lucrative growing jatropha would be.[15]

In 2007, another jatropha processing plant was set up in Mount Hampden, Harare, with the capacity to produce 100 million litres of biodiesel annually. With this project, Zimbabwe had again taken the initiative ahead of all other countries in Africa, as this processing plant was the first such commercial plant on the continent.[16] The plant was jointly owned by the Reserve Bank of Zimbabwe and private investors. The fuel crisis crippling the economy at the time owing to lack of foreign currency in the country was the major driving force that led the government to resuscitate the programme. It was estimated that the project would save the country from diesel importation to the value of close to US$80 million annually. The plant produced over 500,000 litres of biodiesel in 2008 and almost 200,000 litres in 2009, using cotton seed supplied by Cargill and Cottco as feedstock[17]. This was less than 5% of its output capacity.

14 (Ibid.).
15 Mubonderi (2012).
16 Moyo et al. (2014).
17 (Ibid.).

Then in 2009, the multi-currency system was introduced and companies were allowed to import fuel independently. This led to fuel becoming available on the market and hence biodiesel production was no longer seen as a priority. Finally in April 2010, the government stopped funding its biodiesel programme. The main reasons for the project folding was lack of coordination in running the programme and the inability of NOCZIM to pay an attractive price to the farmers for jatropha – which eventually led to lack of constant and reliable supply of the feedstock.[18]

On the bioethanol front, in February 2008, the Ministry of Energy and Power Development (MoEPD) presented in Cabinet a memorandum that proposed the development and use of biofuels in order to mitigate the impact of the rise in fuel prices on the economy. Cabinet approved the principles for the development and use of biofuels in Zimbabwe and invited the Minister to proceed accordingly. In May 2008, Triangle had started to produce hydrous ethanol. Sugarcane production at Triangle is based on the combination estate or 'plantation with out-grower' model. The estate supplies the largest proportion of feedstock. There are about 300 out-growers, each with a 99-year lease agreement on plots ranging in size from 10 to 82 ha. Tongaat (Triangle's parent company) offers technical and input assistance, the value of which is then subtracted when the farmers sells the cane to the miller.

In 2011 a large bioethanol plant project was resuscitated at Chisumbanje in the Chipinge District. The plant was set up on a joint venture partnership between the governmental Agricultural Rural Development Agency (ARDA) and a private investor, Green Fuels. It was set to be the largest of its kind in sub-Saharan Africa. It was envisaged that when fully operational, the plant would produce 700,000 litres of fuel daily equalling 27% of the country's fuel needs, and achieving foreign currency savings in the order of US$100 million per year (depending on international fuel prices). In 2011 Green Fuel did start producing anhydrous ethanol and began selling ethanol on the market the same year. During the third quarter of the year, FMI Energy (BP) began to dispense blended petrol at its retail sites. The bioethanol production facility was expected to create thousands of jobs and lower local fuel prices, and also to improve the livelihood of rural folk living within the vicinity of the plant. Unfortunately, in February 2012, the plant shut down its operation owing to numerous challenges it was facing, including a dispute between the villagers and Green Fuels and the high

18 (Ibid.).

price of the blended fuel. It has since resumed operations, but is still running well below capacity.

In 2013, government made blending of ethanol in unleaded petrol mandatory through Statutory Instrument 17 and Statutory Instrument 147A. Initially, the blend was set at 10%, meaning that all licensed procurers and wholesalers of unleaded fuel had to ensure it had been blended with a minimum of 10% of ethanol produced by a licensed producer. This mandate was raised to 20% in March 2014. However, owing to lack of ethanol supply, it was pushed back to 10% in April 2015

B. Developing a national biofuels policy for Zimbabwe

I. The project: Facilitation of the Development of a National Biofuels Policy of Zimbabwe

The EU-funded project, Facilitation of the Development of a National Biofuels Policy of Zimbabwe (NBPZ), was officially launched in March 2015 to assist the MoEPD in its endeavour to develop a comprehensive, sound and inclusive national biofuel policy by:

- Providing a systematic overview of other countries' experience with biofuels, with a special focus on best-practice case studies;
- providing technical input on the four key pillars of the NBPZ (economic, agricultural, environmental and social and institutional); and
- addressing the legitimate concerns of all the relevant stakeholders with an interest in or likely to be impacted on by a biofuel industry in Zimbabwe.

If implemented rigorously, the NBPZ underpinned by the abovementioned processes should:

- Provide an enabling environment for the development of a biofuel sector in Zimbabwe; and
- ensure that the biofuel production, processing, distribution and marketing in Zimbabwe will remain within the parameters of environmental and social sustainability.

Central to the project were stakeholder engagements for policy development. Gathering input from the various stakeholders was deemed important for the following reasons:

- To legitimise the NBPZ;
- to ensure that all key sectors are part of the process;
- to provide information on purpose, progress and expected outputs of the policy;
- to allow stakeholders to voice their questions and concerns;
- to decipher initial (historical) thinking and vision they have with the re-introduction of biofuels as an alternative fuel;
- to identify gaps and resolve major questions of interpretation;
- to learn from the experiences of the various biofuel initiatives enacted to date to understand 1) what has been done well, 2) what was not done well and why, 3) what are the areas of improvement; and
- to consult a number of targeted stakeholders (technical professionals) to verify information, get additional analytical data, sharpen the focus and improve relevance of issues for NBPZ.

The project was formally concluded by handing the MoEPD a policy document at the end of 2015 that was then subject to Cabinet approval.

II. Policy anchoring

The national biofuel policy must consider and integrate into the country's existing development plans, as well as into any approved sectoral and sub-sectoral policies and strategies.

The current national development blueprint in Zimbabwe is the Zimbabwe Agenda for Sustainable Socio-Economic Transformation (ZIMASSET) document, which covers the period 2013–2018. The intention of ZIMASSET is to achieve sustainable development and social equity anchored on indigenisation, empowerment and employment creation. For the energy sector, ZIMASSET focuses on creating value through processing and beneficiating natural resources for the increased use of alternative forms of energy and reducing diesel and petrol imports. This should have been made possible by land reform, which opened access to land to a large number of Zimbabweans. The promotion of biofuels in the country is still seen as an opportunity for both smallholder and large-scale farmers to participate actively in the biofuels value chain.

Although not much talked about, Zimbabwe, furthermore, has a Vision 2020 document in which "Zimbabwe is to be a united, strong, democratic, prosperous and egalitarian nation with a high quality of life for all Zimbabweans by the year 2020".

In line with these economic blueprints, Zimbabwe adopted a National Energy Policy (NEP) in 2012. Together with the National Energy Policy Implementation Strategy (NEPIS, 2012), these documents outline the country's vision, objectives, strategies and goals for the energy sector. The MoEPD has since included the development of a national biofuels policy in its Five Year Development Plan.

The Vision of the MoEPD, which is reflected in the NEP, is to achieve universal access to sustainable energy in Zimbabwe by 2030. More specifically, the NEP has the following objectives:

- Increase access to affordable energy services for all sectors of the economy through optimal use of available energy resources and diversification of supply options;
- improve the institutional framework and governance in the energy sector to enhance efficiency and energy services delivery;
- promote research and development in the energy sector; and
- develop the use of other renewable sources of energy to complement conventional sources of energy.

Biofuels are part of the renewable energy subsector that the NEP aims to develop and thus the NBPZ will be a subsidiary of the NEP.

Finally, the NBPZ must also be aligned with the country's regional and international cooperation agreements. The two most important regional agreements are the SADC Energy Protocol and the SADC Renewable Energy Strategy and Action Plan, to which Zimbabwe is a signatory.

The SADC Energy Protocol has the following objectives: [19]

- Harmonisation of national and regional energy policies;
- cooperation in the development of energy and energy pooling;
- ensuring the provision of reliable, continued and sustainable energy services in the most efficient and cost-effective manner;
- promoting joint development of human resources and organisational capacity building; and
- cooperation in research, development, adaptation, dissemination and transfer of low-cost energy technologies.

The SADC Renewable Energy Strategy and Action Plan has the following key strategic objectives:

19 SADC (2000).

- Achieving sustainable energy security through reliability and security of supply, particularly of modern energy services;
- increasing access to modern energy services particularly to the poor that largely depend on inefficient traditional forms of energy in order to enhance their socioeconomic status and to meet the Millennium Development Goals (MDGs);
- offsetting the large, hard currency energy import bills resulting from high fuel imports (especially with high oil prices) through the introduction of alternative fuels and energy sources; and
- becoming part of the global effort to seek low carbon development paths and build energy systems that are environmentally sustainable and resilient to the impacts of climate change.

Internationally, Zimbabwe is a signatory to the UN Framework Convention on Climate Change (UNFCCC) and related protocols that define the overall framework for intergovernmental efforts to tackle the challenge posed by climate change.

In line with the policy framework given above, the NBPZ should assist the country in improving its energy security; achieving social equity, creating employment and empowerment and helping reduce poverty in rural areas; improving environmental sustainability of its energy supply; reducing the country's energy import bill; increasing access to modern energy services; and supporting the global effort to reduce emissions of greenhouse gasses (GHG).

III. Policy scope

To focus the policy development process, the scope of the policy was defined along a number of parameters. Just as importantly, where necessary, the scope was not pre-defined in order to allow for the optimal outcome to emerge through the consultative process. The scope parameters included:

- Time horizon: The policy is meant to cover the period up to 2030.
- Type of biofuel: Types were not pre-defined, but availability of conversion technology as well as affordability was considered. In this respect 1st generation biofuels and 2nd generation biofuels (based on conventional conversion technologies using non-food crops) are likely to dominate the biofuel market in the short-to-medium term and this is why the NBPZ is focusing on these two biofuel groups. In addition, it has been recognised that the ZIMASSET document points to bioetha-

nol from sugarcane and biodiesel from jatropha as key priorities until 2018.
- Biofuel application: the NBPZ is to be focused on the use of liquid biofuels in the transport sector.
- Biofuel target: This was not pre-defined; but rather the integration of the existing blending mandate for bioethanol into a wider biofuel target was assessed over the course of the project.
- Biofuel market: This was not pre-defined; but the option to supply the domestic market, as well as produce for export was considered during the policy development process.

The NBPZ is built on four pillars – the economic, agricultural, environmental and social, and institutional pillars, which elaborate on a number of aspects of developing a successful and sustainable domestic biofuel sector. The economic issues that were researched in preparation of the policy are outlined in the following section.

C. Key economic aspects of the biofuel sector in Zimbabwe

I. Biofuels value chain linkages and relationships

The biofuel value chain in Zimbabwe is fairly complex, with a number of private and public stakeholders interacting in more or less equitable relationships. Figure 1 below provides an overview of actors involved in each stage of the value chain, which is a useful introduction to understanding the complexities of the sector.

Figure 1: Actors in the biofuel value chain in Zimbabwe

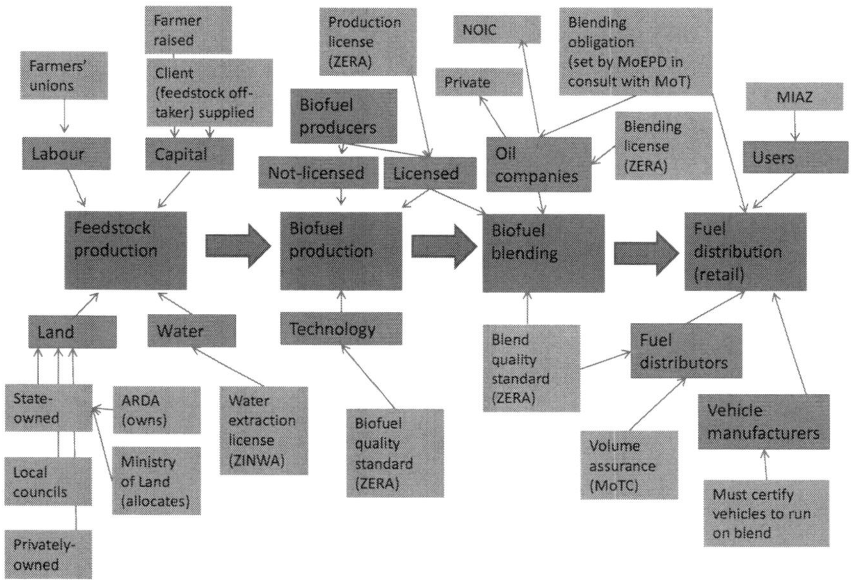

In Zimbabwe, there are concerns mainly in the first two steps of the biofuel chain, namely the feedstock production and the biofuel production or processing stage. Feedstock production is hindered by unresolved land issues, lack of access to capital for investment, infrastructure gaps, etc., but also by the relationships among key stakeholders. For instance, lack of transparency during the sugarcane weighing and sugar content determination process is affecting trust between farmers and ethanol producers. Independent oversight should be established to ensure that fair compensation is given to feedstock producers.

In the biofuel production stage, the current over-reliance on a monopolistic ethanol producer is causing inefficiency in the ethanol market.

The blending stage is currently well organised, with oil companies and fuel distributors having the choice of independent blending or having their fuel blended at the National Oil Infrastructure Company (NOIC) depots at competitive prices. Low blends do not require any additional distribution infrastructure. The wider distribution of a high blend, on the other hand, would require fuel distributors to invest in separate storage and handling capacities and they should be allowed sufficient time to make the necessary investments.

The Zimbabwean Energy Regulatory Agency (ZERA) is overseeing the blending process through its blending licence requirements and regular quality checks of the blends. The Standards Association of Zimbabwe has already produced quality standards for both ethanol and biodiesel that are deemed appropriate and are in line with international standards, allowing Zimbabwean ethanol the option to be exported to international markets. ZERA carries out regular independent quality control checks to ensure that the quality standards are complied with. So far, ethanol produced in Zimbabwe has by and large been found compliant with the applicable quality standards.[20]

The existing transportation, storage and blending infrastructure in the country is deemed appropriate to support expansion of the domestic biofuel market and no major interventions are expected to be required in the short-to-medium term in these segments of the value chain.

At the fuel distribution stage there has been strong opposition to biofuels. This was due to insufficient information on the benefits of introducing biofuels into Zimbabwe's transport fuel mix, but also because of the blending ratio set by the Statutory Instrument 17 of 2013, which mandated a 20% ethanol content in petrol (on a volume basis). This has caused an outcry by motorists, most of whom own vehicles warranted to run on blends of up to 10% only. The country still has very few vehicles capable of running on higher blends, but the main ethanol producer (Green Fuel) has had some success in installing conversion kits on conventional vehicles that allows them to run on high ethanol blends.

II. Economics of biofuel production

The economics of biofuels are determined by their production costs and market price relative to other fuel options. Production costs depend on a number of variables, the most important of which are:

Feedstock production costs: These costs account for approximately half of the cost of ethanol from sugarcane, and for significantly more in the case of the starch-based bioethanol (such as from corn or grain).[21] For bio-

20 Personal communication with ZERA.

21 Bauen et al. (2009).

diesel, this figure can reach as high as 80–90% of total production costs.[22] Feedstock production costs in turn depend on:

- Input costs, namely land, labour, water, energy, and capital costs;
- Seed, fertiliser and farm mechanisation costs;
- Feedstock storage and transportation (to conversion facilities) costs; and
- Plant yield.

In Zimbabwe, several of the cost components mentioned above are high (especially energy and water costs, considering the erratic supply, but also capital costs, fertiliser costs, etc.), while yields are low-to-medium compared to best possible yields. This translates into relatively high prices for feedstock. In Zimbabwe, average sugarcane production cost is some US$700/t,[23] compared to some US$350/t in Brazil, the world's largest sugarcane producer. Jatropha seeds currently fetch a state-determined price of US$150/t, which is comparable to prices in other regions, but is not sufficient to incentivise farmers to plant jatropha widely.

Feedstock conversion efficiency: As shown in Table 1, 1 ton of sugarcane produces about 80 litres of ethanol by direct conversion from sugarcane (by extracting cane juice) or 10 litres via the molasses route[24] and 1 ton of jatropha seeds produces 300 litres of biodiesel.[25] Zimbabwean biofuel producers are largely achieving such internationally benchmarked conversion efficiencies.

---

22 (Ibid.).

23 Based on 65 000 USD/ha cited in Shumba et al. (2011).

24 Personal communications with sugarcane growers during stakeholder workshops. By way of comparison, 1 ton of sweet sorghum produces 55 litres of bioethanol by direct conversion as cited by Woods (2001).

25 Personal communications with jatropha growers during stakeholder workshops.

Table 1: Average feedstock conversion efficiencies

Sugarcane-ethanol	**Jatropha-biodiesel**
From cane juice **1 t -> 80 *l***	1 t -> 300 *l* (assumptions: oil yield is 30%, conversion from oil to biodiesel is roughly 1:1, specific gravity is 0.89 kg/*l*, hence 1 t of seed produces 300 kg of oil, which is cca 300 *l* of biodiesel)
From molasses **1 t -> 10 *l*** **(assumptions: 4% of sugarcane yield is molasses)**	
1 ha -> 7,200 *l*	1 ha -> 900 *l*

Energy inputs: This may account for as much as 20% of operating costs of biofuel plants.[26] Energy costs are known to be high in Zimbabwe, affecting total production costs.
The scale of the plant: Different plant sizes can be used for processing biofuel feedstocks. They include small (in the order of a few thousand litres per day), medium (in the order of tens of thousands of litres per day) and large (in the order of a million litres per day). Economies of scale are important to reduce average costs, so the trend has been towards larger plants with capacities in the order of 200 million litres per year. Medium to large plants operate on a centralised processing model that also sources feedstock from smallholder out-growers. Zimbabwe has biofuel plants falling in all three scale groups, as shown in Table 2.

Table 2: Biofuel production capacity in Zimbabwe

	Company	**Biofuel production capacity (annual)**
Ethanol	Triangle	41 mio l
	Green Fuel	120 mio l
Biodiesel	Finealt	0.3 mio l

The sale or productive use of by-products: The process of converting feedstock to a biofuel often results in several by-products (Table 3), which can be valorised and have a significant impact on the biofuel plant's profitability and hence the price of the main product. To improve competitiveness of Zimbabwe's biofuels, the valorisation of the by-products should be actively promoted.

26 Bauen et al. (2009).

Table 3: Biofuel by-products and their applications

Biofuel	By-products and their application
Ethanol from sugar-cane	Bagasse – can be used for electricity generation
Ethnaol from cassava	Liquid organic fertiliser and DDW (dry distillers waste) – can be used as high-protein animal feed
Biodiesel from jatropha	Presscake – can be used as animal feed (if de-toxified), biogas feedstock, raw material for bio-briquettes, organic fertiliser
Biodiesel from soybean	Hi protein presscake – highly valued animal feed

Both ethanol producers in Zimbabwe are already using bagasse for power generation. Triangle is currently producing 5 MW electricity (in 2015), which might be increased to 20 MW, and Green Fuels is generating 18 MW using 282,495 tonnes of bagasse, and expecting to increase its power output to 93 MW (including the 40 MW supply from the construction of Condo Dam).[27]

The high input costs discussed above translate to relatively high prices for biofuels. A litre of ethanol in Zimbabwe is sold for between US$0.78 and US$0.95 at farm gate,[28] compared with US$0.3 in Brazil and an average import price of US$0.6 per litre (in 2015).[29] Biodiesel produced from jatropha sourced from smallholders at US$150/t costs about US$0.9/l to produce.[30] These prices are substantially higher than both international prices for biofuels and those of imported fossil fuels and will have to be reduced to allow substantial increase in the use of domestically produced biofuel without causing significant financial strain either on the national budget (because of continued subsidisation) or on the consumers (because of higher prices).

III. Setting biofuel targets

Biofuels can be used in pure form or more commonly in fuel blends, as is also the case in Zimbabwe. Blending ratios determine the level of biofuel

27 Personal communication with Green Fuel.
28 Personal communication with ethanol producers at stakeholder workshops.
29 Personal communication with fuel companies in Zimbabwe.
30 Personal communication with Finealt, the parastatal biodiesel producer.

in the national transport fuel mix and are thus at the core of the country's biofuel ambitions.

Ethanol is typically blended with petrol at 2%, 5%, 10% or 15% by volume in what are known as 'low blends' or at 85% or higher in 'high blends'. Such low petrol-ethanol blends are then known as E2, E5, E10, E15, while a typical high blend is E85. Biodiesel can also be blended and used in many different concentrations. Typical low blends are B20 (20% biodiesel, 80% petroleum diesel), B5 (5% biodiesel, 95% petroleum diesel), and B2 (2% biodiesel, 98% petroleum diesel). On the other end of the spectrum is pure biodiesel, or B100.

As long as biofuels are not cost-competitive with mineral fuels, the amount of biofuels that enter a country's transport fuel mix is politically determined. This determination needs to be informed by the amount of biofuels that can be tolerated by the national vehicle fleet, the amount that can be sustainably produced, and the level of support that can be offered. Each of those is briefly elaborated on here as they should inform the proposed biofuel target for Zimbabwe:

1) In terms of what level of blending can be accommodated by the Zimbabwean vehicle fleet in the short-to-medium term, the following needs to be considered:

   - There seems to be no consensus on what the maximum level is of blending in low-blends that can be used in conventional vehicles, with car manufacturers and oil companies generally in favour of a 10% blending wall, while biofuel producers and farming lobbies argue in favour of a higher blending mandate.
   - Most petrol-based light duty vehicles in Zimbabwe's transport sector today are warranted to run on a 10% ethanol blend only; however, a 15% blend should be acceptable for most engines of cars manufactured after 2001.[31] Nevertheless, imposing a blend higher than 10% would effectively nullify most car warranties in Zimbabwe. Flex-fuel vehicles that can run on any combination of petrol and ethanol are still very few.
   - In the case of biodiesel, as long as the vehicle was manufactured after 1993, blends of up to B20 can be used in diesel engines and fuel injection equipment with little impact on operating performance.[32]

31 See http://www.eia.gov/tools/faqs/faq.cfm?id=27&t=10, accessed 20 June 2015.

32 National Renewable Energy Laboratory (2008).

- The vast majority of countries have set blending targets for ethanol at 10% or less on a volume basis, with only a few blending at 20% or higher. In countries where blends above 10% have been introduced, car manufacturers were typically given a period of time to comply with the new fuel regulations and adapt the engines of the cars that were sold on the local market.

2) Producing large amounts of biofuels requires significant amounts of resources. Zimbabwe does have both the natural resources (sufficient available land, water, and suitable climatic conditions) and labour resources (a significant proportion of the workforce still resides in rural areas, as well as highly skilled agricultural experts driving increases in productivity) to produce sufficient amounts of feedstock to feed a domestic biofuel sector. There are, however, significant constraints on capital inputs, both financial and infrastructural. Unreliable energy supply poses major challenges to irrigation and other activities necessary for high-productivity agriculture. Severe liquidity constraints in the economy limit the ability of farmers and even companies to raise capital for investments in modernisation and expansion of agricultural activity. This implies that biofuel feedstock production is likely to take place on land which is currently used to produce agricultural commodities for other markets. This however, may be the salvation, rather than a problem, for those producers, as some of the traditional export markets for Zimbabwean agricultural commodities have collapsed in recent years. Sugar exports for example, have suffered as a consequence of market liberalisation in Europe, which has led to an oversupply of sugar and has caused international sugar prices to plunge. In this context, a domestic biofuel sector is seen as a possible replacement for the loss of sugar export markets.
   Finally, development of a domestic biofuel sector will require significant government support. This is because biofuels are not cost-competitive with fossil fuels in Zimbabwe, as is indeed the case in most countries. With the current low international oil prices said to persist for the short-to-medium term, biofuels are not likely to become price-competitive at the pump unless they continue to benefit from government support. Most countries recognise this and offer some form of support to biofuels. While there is a wide range of possible measures that governments can consider, most will have an impact on the national fiscus. The most common measure used by governments worldwide to support the uptake of biofuels is by means of tax breaks. This is already in place in Zimbabwe, whereby etha-

nol only carries a duty of 5 cents per litre, as opposed to over 60 cents per litre for unleaded petrol.[33] If such a tax exemption is maintained as ethanol reaches higher penetration levels, the impact of the loss of taxes and duties associated with it will be felt by the national fiscus. Considering the very tight situation of Zimbabwe's national budget, there is real concern that a 20% or more fallout of petrol duties and levies could significantly compromise the country's ability to meet its payment commitments.

At the same time, Zimbabwe's current account is suffering an unsustainable deficit, mainly caused by a negative trade balance, of which fuel imports are a major element. Increasing the use of domestically produced biofuels would thus alleviate the pressure on the current account.

To estimate the impacts of different levels of biofuel uptake, the author presents a number of possible biofuel demand scenarios, and their implications for the macroeconomic parameters discussed above. These are presented in Table 4 for ethanol and Table 5 for biodiesel. For the case of biodiesel, the figures presented in Table 5 should be interpreted as upper thresholds, as some of the diesel imported is consumed directly by industry and that would not be subject to blending (but no reliable figures are readily available as to how much).

33 Personal communication with ZERA.

Table 4: Ethanol demand scenarios and macroeconomic impacts estimates

	2015	2020	2025	2030
Projected total petrol demand by volume	**461.7 mio *l***	**555.2 mio *l***	**635.4 mio *l***	**715.6 mio *l***
ETHANOL DEMAND				
Low-demand scenario: 10% blend only (E10), no high blends	**46.2 mio *l***	**55.5 mio *l***	**63.5 mio *l***	**71.6 mio *l***
Petrol displaced	*10%*	*10%*	*10%*	*10%*
Mid-demand scenario: 10% blend (E10) + 2 % high blend (E85) from 2017 & 10% from 2020	**46.2 mio *l***	**118.9 mio *l***	**136.1 mio *l***	**153.3 mio *l***
Petrol displaced	*10%*	*21%*	*21%*	*21%*
High-demand scenario: 10% blend (E10) + 5% high blend (E85) from 2017 & 20% from 2020	**46.2 mio *l***	**182.3 mio *l***	**208.6 mio *l***	**235.0 mio *l***
Petrol displaced	*10%*	*33%*	*33%*	*33%*
MACROECONOMIC IMPACTS				
Low-demand scenario: Approximate fuel import displacement*	32 mio $	39 mio $	44 mio $	50 mio $
Approximate loss of fuel duties to the fiscus**	27 mio $	32 mio $	37 mio $	42 mio $
Mid-demand scenario: Approximate fuel import displacement*	32 mio $	83 mio $	95 mio $	107 mio $
Approximate loss of fuel duties to the fiscus**	29 mio $	67 mio $	77 mio $	89 mio $
High-demand scenario: Approximate fuel import displacement*	32 mio $	128 mio $	164 mio $	165 mio $
Approximate loss of fuel duties to the fiscus**	29 mio $	106 mio $	121 mio $	136 mio $

**In nominal terms, assuming avg price of landed petrol of 70 cents/l; these savings will increase as the international oil price starts rising again.*

***In nominal terms, assuming a continued low taxation of ethanol at only 5 cents/l and unleaded petrol at the current 63 cents/l.*

Table 5: Biodiesel demand scenarios and macroeconomic impacts estimates

	2015	**2020**	**2025**	**2030**
Projected total diesel demand by volume	**1,140.0 mio *l***	**1,282.1 mio *l***	**1,424.3 mio *l***	**1,566.4 mio *l***
BIODIESEL DEMAND				
Low-demand scenario: 2% blend from 2020 and 5% from 2025	**0**	**25.6 mio *l***	**71.2 mio *l***	**78.3 mio *l***
Fossil diesel displaced	*0%*	*2%*	*5%*	*5%*
High-demand scenario: 2% blend from 2018 and 5% from 2020	**0**	**64.1 mio *l***	**71.2 mio *l***	**78.3 mio *l***
Fossil diesel displaced	*0%*	*5%*	*5%*	*5%*
MACROECONOMIC IMPACTS				
Low-demand scenario: Approximate fuel import displacement*	0	17.4 mio $	48.4 mio $	53.3 mio $
Approximate loss of fuel duties to the fiscus**	0	13.1 mio $	36.3 mio $	39.9 mio $
High-demand scenario: Approximate fuel import displacement*	0	43.6 mio $	48.4 mio $	53.3 mio $
Approximate loss of fuel duties to the fiscus**	0	32.7 mio $	36.3 mio $	39.9 mio $

**In nominal terms, assuming avg price of landed diesel of 68 cents/l; these savings will increase as the international oil price starts rising again.*

***In nominal terms, assuming a continued low taxation of biodiesel at only 5 cents/l and mineral diesel at the current 46 cents/l.*

As expected, savings of fuel imports, as well as fiscus revenue loss, increase proportionally with increasing biofuel penetration in the transport fuel mix. And while displacing increasing amounts of fuel imports has a positive impact on the nation's trade balance, ultimately the level of revenue loss that can be sustained by the national budget will determine the level of biofuels in Zimbabwe's transport fuel mix in the long run (if current support levels continue to be needed).

The country already has sufficient installed ethanol production capacity to meet ethanol demand under even the high-demand scenario, as well as sufficient feedstock (on a national level). The status of the biodiesel sector, on the other hand, is mainly in its demonstration phase, where important progress is being made with regard to jatropha domestication, however the scales are still too small to feed a commercial-sized operation.

Therefore, for the short-to-medium term up to 2030, the following strategy is recommended:

- For ethanol, the ceiling for mandatory low-blends should not exceed 10% biofuel blending in the short-term, so as not to disadvantage motorists whose vehicles are only warranted to run on this level of blend. At the same time, government should engage with car manufacturers and their representatives in Zimbabwe to raise the blending ceiling of conventional vehicles to 15% and still be warranted. Vehicle manufacturers should be given a set time period to adjust their warranties, but should be incentivised to do so as a matter of urgency.

At the same time, limiting the market only to low blends will severely constrain sector development, as the ceiling for a low blends (be it 15% or even 20%) can be achieved fairly easily (provided sufficient feedstock supply is assured). To provide room for further sector development, it is therefore important to support the development of a voluntary high-blend market (E85). This combination of low- and high-blend markets takes into consideration the interest of motorists, while allowing the ethanol sector to grow.

- For biodiesel, a target 2% blend starting in 2020 is proposed, which could be raised to 5% by 2025, if feedstock production is deemed able to keep up with demand without compromising food security. The use of recycled vegetable oils as feedstock for biodiesel should be actively encouraged to alleviate pressure on land use for jatropha production.

IV. Fuel economy and pricing of the blend

It is well-established that the fuel economy of traditional biofuel blends is lower compared to unblended fossil fuels. This is because of the biofuels' lower energy density. While some novel biofuels with superior properties are emerging (for instance biobuthanol), it has to be recognised that most of the biofuels consumed in Zimbabwe in the short-to-medium term will be traditional biofuels with lower energy densities. Ethanol's average energy density is only some 66% that of petrol,[34] while for biodiesel this figure is 93%.[35]

Research carried out by the Scientific and Industrial Research and Development Centre (SIRDC) of Zimbabwe shows that when compared with unleaded petrol, the average fuel economy (defined as kilometres travelled with a litre of fuel) loss is 4.84% with E10, 7.24% with E15 and 11.06% with E20 for local travel. For highway travel, the fuel economy reductions were 4.33%, 7.60% and 19.98% for E10, E15 and E20, respectively.[36] While higher blends have not been tested in the local context, E85, as a general rule, has a fuel economy that is about 15–20% lower compared to pure unleaded petrol.[37] The reductions in fuel economy for low blends recorded in Zimbabwe are somewhat higher than that reported in other parts of the world. Both in the United States and Australia, for example, a 10% blend is generally reported to reduce fuel economy by only 3%.[38]

Considering local prices for biofuel blends, this translates to increased costs for motorists. Based on fuel prices in December 2014, the cost of local travel per km on different blends has been calculated to be $0.117 per km for pure unleaded petrol, $0.131 km for E5, $0.132 for E10 and $0.137 for E20, as presented in Table 6. This represents a relative increase of 11.9%, 12.8% and 17.1% compared to pure unleaded petrol for E10, E15 and E20 blends, respectively. For highway travel, the percentage cost increase for using E10 is only some 1.5%; however, for E15 and E20 it is 10.0% and 18.5%, respectively. Ideally, these additional costs should be

34 See http://www.world-petroleum.org/technology/202-alternative-transport-fuels-courtesy-of-aip, accessed 20 June 2015.

35 See http://www.afdc.energy.gov/fuels/fuel_comparison_chart.pdf, accessed 24 June 2015.

36 SIRDC (2014).

37 See http://www.afdc.energy.gov/fuels/fuel_comparison_chart.pdf, accessed 24 June 2015.

38 See http://www.eia.gov/tools/faqs/faq.cfm?id=27&t=10, accessed 20 June 2015; and http://www.resourcesandenergy.nsw.gov.au/energy-consumers/sustainable-energy/office-of-biofuels/e10-fuel-economy, accessed 20 June 2015.

incorporated in the pricing formula for the blend to avoid an additional burden on the motorists.

Table 6: Cost of fuel blends in Zimbabwe in December 2014 and related travel costs[39]

Fuel blend	Cost per litre (US$)	Cost per km (US$) for local travel	Cost differential relative to E0 for local travel	Cost per km (US$) for highway travel	Cost differential relative to E0 for highway travel
E0	1.46	0.117		0.130	
E10	1.43	0.131	11.9%	0.132	1.5%
E15	1.41	0.132	12.8%	0.143	10.0%
E20	1.39	0.137	17.1%	0.154	18.5%

To maintain fairness, the price of the blend should in principle be commensurate to the reduction in the fuel economy compared to pure unleaded petrol. However, as the reduction in fuel economy of E10 is not material and no pure unleaded petrol should be sold anymore in Zimbabwe (except to users that have successfully applied for an exemption from the blending mandate), this argument falls away and, to avoid an additional administrative burden, E10 could continue to be priced as outlined in the Petroleum Pricing Regulation under Statutory Instrument 80 of 2014. The said regulation should be expanded to accommodate pricing of a low biodiesel blend (B2 and B5) when these are introduced.

With regard to E15 and E20, as SIRDC data shows, even though their price at the pump is lower compared to that of unleaded petrol, the more significant loss of fuel economy results in higher cost per km travelled, both for local and highway travel. If these blends continue to be sold in Zimbabwe, the loss of fuel economy should be reflected in their price.

Similarly, the difference in fuel economy of high blends (E85) is more significant, and should be accounted for in the pricing structure. Therefore, the cost of E85 at the pump should not exceed 85% of the price of pure unleaded petrol.

39 Source: SIRDC (2014) and own calculations based on SIRDC (2014) data.

V. Biofuel incentives and market development

As long as biofuels are not price competitive with fossil fuels, they will need some sort of policy support to be accepted by users. There is a wide array of policy support measures available to policy makers, with the three major categories being:

- Budgetary support measures, either as government equity investment or tax concessions for biofuel producers (refineries), retailers and users, or as direct support to biomass supply, biofuel production capacities, output, blending, specific infrastructure and equipment for biofuel users;
- blending or the use of mandates that require biofuels to represent a minimum share or quantity in the transport fuel market. While these measures generally are neutral for public budgets, the higher production costs of biofuels usually result in increased fuel prices for the final consumer; and
- trade restrictions, mainly in the form of import tariffs, aimed at protecting the developing domestic biofuel industry. At the same time, shielding the less cost-efficient domestic biofuel industry from competition from lower-cost foreign suppliers may result in higher domestic biofuel prices.

Often a combination of the above-mentioned measures is employed to achieve the optimal results. To provide a guaranteed market for producers, blending has already been mandated for ethanol by the Statutory Instrument 17 of 2013, which requires all petrol sold in Zimbabwe to be blended with 20% ethanol (on a volume basis). While a push back in blending obligation to 10–15% is proposed here, an aggressive development of a high-blend market, which could soon exceed demand of ethanol for the low blend, is also proposed.

While significantly more work needs to be undertaken to fully understand all their possible impacts, the following measures should be investigated as tools to support the development of a high-blend market:

- Duty-free importation of flex-fuel cars;
- support for distribution and installation of flex-fuel converter kits;
- shifting of government vehicle fleets to E85 as a demonstration tool; and
- capacity-building among motorists on the benefits of E85.

For the case of biodiesel, ZIMASSET currently stipulates that 300,000 litres of biodiesel from jatropha should be produced per year. While this is a good start, a low blending mandate allowing biodiesel from other feedstocks will also provide guaranteed demand up to a level that can be sustained by domestic feedstock without compromising food security. To clarify conditions for biodiesel production in Zimbabwe, biodiesel licensing conditions similar to those for the production of ethanol should be drawn up soon.

Finally, owing to their higher production costs, biofuels need direct support to be acceptable to consumers at the pump. Continued lower taxation of biofuels compared to fossil fuels, insofar as allowed by the national budget, is an appropriate measure.

There are also a number of budget-neutral incentives that can be considered, especially in terms of improving sector efficiency. Even though there is currently sufficient installed biofuel production capacity in Zimbabwe to meet ethanol demand for both E10 and E85 under the high-demand scenario, as well as to meet a low-blend mandate for biodiesel (B2 or B5) (see Tables 4 and 5), other investors should be allowed to enter the market. This would ensure continuous biofuel supply; and stimulate efficiency within the sector for the benefit of consumers and competition for feedstock for the benefit of farmers.

For the ethanol market, this can be achieved by simply revising the licensing conditions for ethanol production, in particular:

- General Condition 1 requiring that major shareholding (51%) should be held by Zimbabweans; and
- special Condition 11 granting exclusive rights to produce anhydrous ethanol for sale under the national blending mandate to a single producer.

The amendment of General Condition 1 has been made possible by the recent announcement of the forthcoming amendment of the Indigenisation and Empowerment Act that will allow line ministries the discretion to determine the appropriate level of indigenisation on a case-by-case basis. In light of this, the MoEPD will be able to negotiate this aspect with other existing and potential ethanol producers in the country. By opening the biofuel market to other companies, oil companies and other fuel distributors will be able to negotiate biofuel prices based on competitive supply and not a monopoly position of the current single licensed producer. In principle, this should also open the possibility of having a decentralised model of small-scale biofuel plants located in the immediate vicinity of the feedstock, and possibly (co)owned by feedstock producers that would benefit from the value added of the biofuel as opposed to the limited value of the raw material, or at least provide feedstock growers with a wider choice of off-takers. In practice, however, constraints on credit availability, logistical and contractual issues and, of course, the necessary economies of scale might still imply a highly concentrated sector in the short-to-medium term.

In any event, considering the limited domestic demand, additional investment in biofuel production is only likely if exporting biofuel from Zimbabwe is a possibility. To position itself as a biofuel hub, Zimbabwe will have to support the competitiveness of its biofuel producers. This might require government support to lower operational costs of the biofuel producers, such as (again, insofar as the constrained national budget allows):

- Investment in infrastructure;
- duty-free equipment importation;
- tax exemptions; and
- continued government support for R&D for improved feedstock yields.

Bibliography

African Development Bank, 2013, Zimbabwe country brief 2013–2015, at http://www.afdb.org/fileadmin/uploads/afdb/Documents/Project-and-Operations/2013-2015_-_Zimbabwe_-_Country_Brief.pdf, accessed 16 May 2015.

Bauen, A, G Berndes, M Junginger, M Londo & F Vuille, 2009, *Bioenergy – a sustainable and reliable energy source*, IEA Bioenergy, at http://www.seai.ie/Renewables/Bioenergy/Bioenergy_–_a_Sustainable_and_Reliable_Energy_Source_MAIN_REPORT.pdf, accessed 20 October 2016.

Monyau MM & A Bandara, 2015, Zimbabwe, African Economic Outlook, at http://www.africaneconomicoutlook.org/en/country-notes/zimbabwe/, accessed 8 November 2016.

Moyo, P, M Moyo, D Dube & O Rusinga, 2014, Biofuel policy as a key driver for sustainable development in the biofuel sector: the missing ingredient in Zimbabwe's biofuel pursuit, *Modern Applied Science* 8, 1.

Mubonderi, J, 2012, Jatropha: the broom of poverty; myth or reality? A critical analysis of the Zimbabwean jatropha programme in Mutoko district, Practical Action Consulting Working Brief at http://practicalaction.org/media/view/31874, accessed 20 April 2015.

National Renewable Energy Laboratory, 2008, *Clean cities fact sheet – biodiesel blends*, US Department of Energy, at http://www.biodiesel.org/docs/ffs-basics/clean-cities-fact-sheet-biodiesel-blends-april-2008.pdf?sfvrsn=4, accessed 20 June 2015.

SADC / Southern African Development Community, 2000, *Regional indicative strategic development plan*, at http://www.sadc.int/documents-publications/show/Regional_Indicative_Strategic_Development_Plan.pdf, accessed 6 April 2015.

SIRDC / Scientific and Industrial Research and Development Centre, 2014, Fuel economy study; comparing performance and cost of various ethanol blends and gasoline, SIRDC.

Shumba, E, P Roberntz & M Kuona, 2011, *Assessment of sugarcane out-growers schemes for biofuels production in Zambia and Zimbabwe*, WWF Zimbabwe, at http://d2ouvy59p0dg6k.cloudfront.net/downloads/assessment_of_sugarcane_outgrower_schemes_for_bio_fuel_production_2.pdf, accessed 10 October 2016.

Woods, J, 2001, The potential for energy production using sweet sorghum in southern Africa, *Energy for Sustainable Development* 5, 1.

The impact of the politicisation of the biofuels sector in Zimbabwe*

Vincent Musewe, BSc (Economics), University of Zimbabwe, Zimbabwe

A. Introduction

As global warming continues to be a challenge, especially in sub-Saharan Africa,[1] there is a need to look into fossil fuel alternatives for sustainable energy supplies. Zimbabwe has not been spared the negative effects of climate change which have not only affected the food and energy sectors, but also threaten the livelihoods of vulnerable communities. A deadly combination of extreme weather conditions and rampant 'anti-nature' behaviour, mainly associated with the blatant disregard for the rule of law; repeated violations of property rights; and pervasive shunning of constitutionalism have worsened the lives of most Zimbabweans.

Unfortunately, the pursuit of alternatives for sustainable energy supplies has been met with a number of challenges, most of which can be traced to undue political interference. Unlike most countries in the Southern African Development Community (SADC) region, Zimbabwe's energy crisis is exacerbated by the inability of the public power entity, Zimbabwe Electricity Supply Authority (ZESA), to meet the country's energy demand. Zimbabwe's electricity consumption at peak levels is about 1,200 MW, even at a time when the country's industry capacity utilisation is at 35%. However, owing to years of incompetent management, lack of capitalisation and corruption, ZESA has totally failed to invest in efficient electricity power production.

This dysfunctional power utility has managed to supply just under 800 MW at any one time, thus subjecting the country to debilitating power outages. The ruling Zimbabwe African National Union Patriotic Front (ZANU-PF) government has, since the mid-1980s, been talking about "mega power deals", but nothing tangible has been achieved.

---

* Special thanks shall be rendered to Mr Rejoice Ngwenya, COMALISO Zimbabwe for his valuable input regarding this chapter.

1 See Niang & Ruppel (2014).

Expanding commercial, domestic, academic, mining and agricultural sectors has exerted so much pressure on energy demand that the pursuit of alternative sustainable energy supplies has become a matter of urgency. It is certainly not sustainable for the country to survive on excess power handouts from Mozambique's Cahora Bassa hydroelectric generation station and South Africa's ESKOM.

In an effort to be energy independent, Zimbabwe has over decades looked into the possibility of utilising its vast biomass resources to produce biofuels. The potential for biofuel development to create jobs, foster rural development and enable energy sufficiency and security is considerable, provided that the country is able to implement its policies effectively.

The production and use of biofuels has been part of Zimbabwe's energy strategy from as early as the 1970s. The jatropha plant (*Jatropha curcas L.*) was used in the production of fuels in the 1970s and 1980s, after being introduced into the country in the 1940s to be exploitation for soap and lubricant production. During the Unilateral Declaration of Independence (UDI) era, the Rhodesian government made notable advances in producing a fuel blend from ethanol.[2]

In 1980, Zimbabwe started producing bioethanol using molasses as the feedstock at its Triangle Ethanol Plant. This was the first production facility of its kind in Africa. The bioethanol plant was producing about 120,000 litres per day and 40 million litres per year. Petroleum blending was done at a rate of 12–15% (v/v) with ethanol. However, the production plant ceased operation in 1992 owing to severe drought conditions, which led to failure to access the feedstock, while at the same time unblended fuel also became cheaper.

In 2005, the country once again relooked into the prospect of using the jatropha plant for biodiesel production. The National Bio-Diesel Feedstock Production Programme was launched with the aim of establishing jatropha plantations right across the country's 10 provinces to ensure sustained supply of the feedstock. It was estimated that the project would save the country close to US$80 million annually from diesel importation. The National Oil Company of Zimbabwe (NOCZIM), which has since

2 The United Nations imposed economic sanctions on Rhodesia in 1965 that made it difficult for the Ian Smith-led government to import fossil fuels. By 1978, Mobil was attempting to advance the production of petrol from methanol in order to promote fuel import substitution and mitigate the 1973 oil crisis. See also Hollaway (1985).

collapsed, was mandated to spearhead the project. This project was an extension of the national energy policy, which bragged:

> jatropha production will improve Zimbabwe's energy security by annually contributing 360 000 tons of Jatropha curcas seeds for processing into 110 million litres of biodiesel by 2010, to provide import substitution and thereby save scarce foreign currency of about USD110 million per annum by 2010. It will create employment opportunities for Zimbabweans in research, production, processing and retailing of biodiesel, empower rural households and alleviate poverty through the growing of Jatropha as a cash crop, reclaim and rehabilitate thousands of hectares of degraded lands through growing of Jatropha as well as providing cost effective farm boundaries.

A US$6 million jatropha processing plant was set up in 2007 in Mount Hampden, Harare, with the capacity to produce 100 million litres of biodiesel annually. The plant was jointly owned by the Reserve Bank of Zimbabwe (RBZ) and private investors. Owing to poor planning, the plant operated at under 5% capacity. This was attributed mainly to a severe shortage of the feedstock. In April 2010, the government stopped funding the plant and that was the end of that.

In 2011 the country saw yet another biofuels initiative, where a US$600 million bioethanol plant project was resuscitated at Chisumbanje in the Chipinge District. The plant was set up as a joint venture partnership between the government, the Agricultural Rural Development Agency (ARDA), and a private investor, Green Fuels Pvt Ltd. It was set to be the largest of its kind in sub-Saharan Africa. The project was initially set to run on the build–operate–transfer model with the government set to take full control of the plant in 30 years' time from Green Fuels. When fully operational, it was expected that the plant would produce around 700,000 litres of fuel daily (27% of the country's fuel needs), saving the country close to US$200,000 per day on fuel imports. The bioethanol production facility was expected to create 4,500 jobs, reduce local fuel prices, and also improve the livelihood of rural folk living in the vicinity of the plant. Once again, on 6 February 2012, the plant shut down its operation citing numerous challenges which it was facing, putting an estimated 4,500 workers at risk of losing their jobs.

Some of the problems experienced include failure to secure a ready market for ethanol, as government had not made fuel blending mandatory. This consequently led to a situation where there was an accumulation of ethanol stock at the plant as it was not selling. In addition to the poor sales, the logistical and infrastructure for blending in the country were reported as inadequate. Other challenges that pushed the plant to shut down its operations were a dispute between the villagers and Green Fuels,[3] and the high price of the blended fuel.[4]

B. Poor policy implementation

In 2011 the Zimbabwean government proposed a new biofuels policy which sought to promote the use of sorghum and oilseeds as biofuels feedstocks, in addition to jatropha and sugarcane. The overall objective of Zimbabwe's biofuels programme is to produce at least 10% of the country's liquid fuel requirements by 2017. The policy on biofuels, once approved, is expected to lay the framework that will regulate and promote investment, production, marketing, and the use of biofuels. However, 2016 is already at an end and the policy has not been finalised yet.

Policy implementation remains an albatross impeding Africa's development. Many African governments remain bad implementers of apparently good policies, and Zimbabwe is no exception. For example, in 1981 Zimbabwe saw the introduction of Growth with Equity, then the three-year Transitional National Development Plan. The first five-year National Development Plan (1986–90) policy was intent on economic transformation and growth, and, by 1990, the country had adopted the Economic Structural Adjustment Programme. The Framework for Economic Reform followed in 1991, before ZIMPREST 1996–2000 geared for a stable macroeconomic environment. The new millennium came with the Millennium Economic Recovery Programme (MERP) in the shadow of the highly partisan Fast Track Land Reform Programme that decimated commercial farming property rights and sent Zimbabwe's economic fortunes headlong into turmoil.

ZANU-PF introduced the Ten Point Plan followed by MERP in 2002 under the Post-Election Economic Development Strategy, then the National Economic Revival Programme in 2003, while the Macro-economic Policy Framework (MEPF) was introduced in 2005, and the National

3 Claris Madhuku, Director of Platform for the Youth Development Trust, has been at the forefront of advocating for the rights of the Chisumbanje and Chinyamukwakwa villagers who are aggrieved in that Green Fuels compromised their traditional property rights. See also Sapp (2015a and b).

4 During the Government of National Unity, the then opposition parliamentarian and energy minister Elton Mangoma led a campaign of refusal to accept Green Fuels' pricing model on the basis that it was meant to benefit Billy Rautenbach, the ZANU.PF crony main shareholder of that company.

Economic Development Priority Programme and Zimbabwe Economic Development Strategy (NEDPP ZEDS) aborted at conception in 2008. The Short Term Recovery Plans 1 and 2 were post-Government of National Unity blueprints designed to deal with broad developmental and growth policies. In a desperate attempt to arrest worsening economic conditions, President Mugabe introduced the 10-Point Plan for Economic Growth in 2013, before following it up with the current, disastrously managed Zimbabwe Agenda for Sustainable Socioeconomic Transformation (ZimAsset).

On its part, the Ministry of Energy and Power Development published the Zimbabwe National Energy Policy in 2012. The policy provides 'interim' guidelines on liquid biofuels, such as ethanol/petroleum blending targets of 20% by 2015, and 5% for biodiesel by 2020. The policy also highlights the need to use non-food feedstock, promote out-grower schemes, and conduct research on improving feedstock yields and processing technology. The policy unfortunately fails to address a major critical point concerning production, which is the land issue.

A well-formulated biofuel policy needs to address two critical issues, namely the production of the biofuels (supply), and the laws governing their use (market and demand). In addition, cross-cutting issues may be addressed, which are directly or indirectly affected by either the production or utilisation of biofuels, or both, within the borders of Zimbabwe.

The policy on biofuels, once approved, is expected to lay the framework that will regulate and promote investment, production, marketing, and the use of biofuels.

C. Implications of predatory politics

Predatory states have existed for a very long time and Zimbabwe is one such example. The key features of predatory rule are: extreme autocratic rule; personality politics and the use of violence, fear and repression; erosion of formal public institutions; pervasive corruption; a disregard for traditional, customary or informal institutions; and rewards for key collaborators, leaders and associates who use power for personal ends and establish a predatory coalition. Key to a predatory state is a predatory coalition which props up a predatory leadership, rent-seeking and economic benefits through favours granted by the state. This has been the case in Zimbabwe's biofuels sector.

Biofuels production requires access to large tracts of land, a sector that is mired in controversy. The government of Zimbabwe – keen to satisfy the needs of few private biofuel investors prepared to sidestep international concerns about the country's political problems – has tended to favour a predatory coalition at the expense of community needs. Chisumbanje therefore clearly represents predatory politics at play. In this part of Zimbabwe, the allocation of land and contracts has been driven mainly by political interests, without taking into account the potential adverse socioeconomic impacts. This is where the development of the biggest bioethanol plant in Africa resulted in the dispossession of local people from their farmland and other land-based livelihood activities.

Chisumbanje is a village in the province of Manicaland, Zimbabwe. It is located in the Dowoyo communal land on the eastern bank of the Save River, about 95 km south of Birchenough Bridge on the Birchenough Bridge-Chiredzi road. The village is in Chipinge District and is bordered by villages such as Chinyamukwakwa and Mashubi.

The Chisumbanje bioethanol plant development started in 2009 and is claimed to be the largest in Africa. The ethanol plant was a US$600 million project commissioned in 2010. After operations began, it was producing 70,000 litres of fuel per month by August 2011. The plant has over 5,000 hectares of land under sugarcane to sustain its production levels. There are also around 400 out-growers in surrounding communities. The project is a joint partnership entered into by the Agricultural and Rural Development Authority with Madcom Rating, Green Fuel Investments and Madcom Investments.

The main funder of these private firms is Billy Rautenbach,[5] a controversial businessman, who is on European Union and United States targeted sanctions lists because of his close association with the ruling party ZANU-PF. Rautenbach's 'unorthodox' business practices have earned him notoriety status within Zimbabwe's opposition circles.

The Chisumbanje biofuel project represents a unique case, as the state has engineered and supported the forced removal of large numbers of rural households from what residents consider to be *their* communal farmland – an act contrary to its Fast Track Land Reform Programme targeting and forcibly expropriating mostly white-owned commercial farms for the pur-

5 In 2004, South African authorities levelled tax evasion charges on him leading to a 50 million rand fine. He also has interests in transport and mining in the Democratic Republic of Congo. His disinvestment in the Hyundai assembly project in Botswana was also mired in controversy. See https://en.wikipedia.org/wiki/Billy_Rautenbach, accessed 19 June 2016.

poses of land redistribution to poor black farmers. Among the fundamental concerns most reported by local users was the way their communally owned land was acquired by private investors through state intervention.

The key problems which arose included:

- The project resulted in severe poverty and lack of alternative opportunity for many families who were displaced. As of June 2012, figures from the Chipinge Rural District Council indicated that out of the 1,733 families displaced by the company, only 499 were each allocated land. The displaced households also experienced intimidation and the disrespectful attitude of the private investor management firm, supported by state agencies.[6]
- Women within households are responsible for food provision, and loss of land meant that they faced serious challenges in feeding their children. Green Fuels is alleged to have found ways to use women for cheap social reproduction, thus actively exacerbating gender inequality existing in society.
- Local communities lost most of their land to pave the way for sugarcane plantations for the biofuel project. There was an overall decline in the standard of living of communities compared to their situation prior to the arrival of the plant. The loss of land meant the loss of important cash income from cotton farming.
- The loss of access to livestock fodder also occurred as a consequence of grazing land being taken over by the project. Prior to the arrival of the biofuel company, grazing across communal lands was a common and important practice, but now no longer possible.
- Water and air pollution caused by the ethanol plant was a frequently cited concern among the community. There were claims that the company discharged untreated, fertiliser-rich waste water from the plant directly into open water sources used by local communities for washing clothes and bathing, and for livestock care.

---

6 Green Fuels was reported as having roped in the Joint Operation Command (JOC) to suppress villagers who were demanding that the ethanol-producing company fulfil its social community responsibility obligations. JOC comprises the provincial heads of the Central Intelligence Organisation, Zimbabwe National Army, Zimbabwe Republic Police and Zimbabwe Prison and Correctional Services. Villagers were promised two hectares of irrigated land but received 0.5 hectares. Their livestock are routinely impounded for straying into the company's sugarcane plantations.

Clearly for many reasons, the Chisumbanje biofuel investment has failed to live up to its expectations and promises. Instead of generating sustainable benefits, biofuel development has contributed to the loss of livelihood sources (land and natural resources), indicating a clear collision between national interests and local livelihood needs.

What can be learned from the case above is that transforming predatory rule into a political order that is more developmental and that can address issues of growth, stability and poverty reduction is key if Zimbabwe is to benefit from its stated energy objectives.

D. *New biofuels policy*

With input from relevant stakeholders, the World Wildlife Foundation facilitated the development of a national biofuel policy framework that was included in its 2011 annual report. The framework proposed four policy pillars namely economic, social, environmental and institutional.

A decision by the Ministry of Energy and Power Development to prioritise national biofuel policy development in its five-year development plan for 2013 has given impetus to this project component. The policy development process will be spearheaded by the Ministry with the full participation by the Ministries of Agriculture, Mechanisation and Irrigation Development, and the Ministries of Environment, Water and Climate. It will take about 12 months and will involve the following key steps: the development and approval of terms of reference for a consultant to draft the policy based on a framework developed by the Project; the engagement of a consultant to produce a draft policy; the conducting of national, provincial and district level stakeholder consultations on the draft policy; and the review and approval of the draft policy by relevant arms of government. If the process is indeed as participatory as it sounds, one hopes it will exorcise and de-toxify the unpleasant politics in the biofuels sector.

E. *Conclusions and a way forward*

The future of biofuels in Zimbabwe is not necessarily bright – for the reasons articulated above, and also because of the prospects of cheaper fossil fuels. Moreover:

> [T]hose who oppose biofuels argue that biofuels production will cause food insecurity around the world. The UN-Energy stated that rapid growth in liquid fuels production will raise agricultural commodity prices and could have negative eco-

> nomic and social effect particularly on the poor who spend a large part of their income on food.[7]

In Mugabe's Zimbabwe where chaotic land reform is already a food security risk, dangling a biofuels feedstock carrot at villagers will only exacerbate the situation. Food shortages may cause high food prices as more land and water are dedicated to more profitable biofuels production than food production.

Another area of concern is the coercive politics of blending. Apart from the fact that patronage, rent-seeking and bully tactics are the modus operandi of ZANU-PF governance, Zimbabwe Energy Regulatory Authority chief executive Gloria Mgombo's (mandatory blending of anhydrous ethanol with unleaded petrol) (Amendment) regulations, published in the Statutory Instrument 147A of 2013, adopt a very high coercive and obligatory stance. The typical 'this-is-a-good-policy-take-it-or-else' ZANU-PF approach does not augur well for virtues of participatory national governance.

Most enlightened Zimbabweans concur on benefits of ethanol blending, which include, among others, reduction in vehicle tailpipe emissions, improvement in the octane rating of fuel, and energy security for the country, but the message, dispensed with a political whip, tends to be bitter-sweet. It would not help the situation if the country were embroiled in another case of litigation, as in the case where rights activist and advocate Thabani Mpofu sued former Energy and Power Development ZANU-PF Minister Dzikamayi Mavhaire, the Zimbabwe Electricity Regulatory Authority and Green Fuels over Green Fuels' monopoly and blending ratios to blending levels beyond E10 petrol blending.

Given the extent of the conflict as expressed by Claris Madhuku, director of Platform for Youth Development Trust, I am not convinced that the government has put in place a fool-proof and universally accepted mechanism on community livelihoods, land tenure systems, and water supply. Just like the Finalt (Pvt) Ltd company jatropha project in Mutoko, communities still argue that not even out-growing takes into account their concerns. Says Madhuku on his Face Book post:[8]

> PYD has for the past 7 years interacted with many researchers and government officials where facts have been interrogated. Chipinge South and in particular Chisumbanje and Chinyamukwakwa have recorded bumper harvests in cotton, maize

7 UN-Energy (2007).

8 See https://www.facebook.com/claris.madhuku?fref=nf, accessed 10 June 2016. See also Makore (2010).

> and sorghum before Green fuel (sic). Statistics held by the Ministry of Agriculture ftp://ftp.fao.org/docrep/fao/010/a1094e/a1094e00.pdf confirms this with known farmers like Minyizeya of Matikwa and Muchatiroto Mashava of Chinyamukwakwa as living examples. PYD promises to stand strong with the community to challenge false beliefs that life before Green fuel was worse and that only Green fuel has provided life.

If only the new energy policy could neutralise the negative political vibe in the biofuels sector, then Zimbabwe can indeed be a fully integrated member of the sustainable energy community. The jury is still out on that one.

Bibliography

Holloway, J, 1985, Rhodesia Search for Synthetic Fuels, *New Scientist*, 21 February 1985, No 1444, 34 at https://books.google.co.zw/books?id=slEuRovCc9cC&pg=PA35&lpg=PA35&dq=did+Rhodesia+produce+any+ethanol?&source=bl&ots=H6iCGEXUS6&sig=b-BeTenpFE27ViuLclO_6TIy22g&hl=en&sa=X&ved=0ahUKEwi-zPqStrPNAh-WGuRoKHVqfCyIQ6AEIHDAA#v=onepage&q=did%20Rhodesia%20produce%20any%20ethanol%3F&f=false, accessed 19 June 2016.

Makore, G, 2010, Community participation in biofuels crop production in Zimbabwe: A focus on the policy and practical-aspects, Harare, Zimbabwe Environmental Law Association, SIEMENPUU, at http://www.slideshare.net/GilbertMakore/community-participation-in-biofuels-crop-production-in-zimbabwe-a-focus-on-the-policy-and-practical-aspects/, accessed 19 June 2016.

Niang, I, OC Ruppel, 2014, "Africa", in: Field, CB, VR Barros, DJ Dokken, KJ Mach, MD Mastrandrea, TE Bilir, M Chatterjee, KL Ebi, YO Estrada, RC Genova, B Girma, ES Kissel, AN Levy, S MacCracken, PR Mastrandrea, LL White (eds), *Climate Change 2014: Impacts, Adaptation, and Vulnerability – Contribution of Working Group II to the Fifth Assessment Report of the Intergovernmental Panel on Climate Change*, Cambridge, Cambridge University Press, 1199.

Sapp, M, 2015a, Workers and cane growers strike in Zimbabwe as Green Fuel fails to pay, *Biofuels Digest*, 12 May 2015, at http://www.biofuelsdigest.com/bdigest/2015/05/12/workers-and-cane-growers-strike-in-zimbabwe-as-green-fuel-fails-to-pay/, accessed 19 June 2016.

Sapp, M, 2015b, Zimbabwe parliament wants Green Fuel to compensate villagers for alleged pollution, *Biofuels Digest*, 23 February 2015, at http://www.biofuelsdigest.com/bdigest/2015/02/23/zimbabwe-parliament-wants-green-fuel-to-compensate-villagers-for-alleged-pollution/, accessed 19 June 2016.

UN-Energy, 2007, Sustainable bioenergy: A framework for decision makers, at ftp://ftp.fao.org/docrep/fao/010/a1094e/a1094e00.pdf, accessed 9 December 2016.